# SPACELAB

## RESEARCH IN EARTH ORBIT

David Shapland and Michael Rycroft

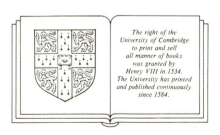

The right of the
University of Cambridge
to print and sell
all manner of books
was granted by
Henry VIII in 1534.
The University has printed
and published continuously
since 1584.

## Cambridge University Press

Cambridge
London    New York    New Rochelle
Melbourne    Sydney

# For everyone interested in Space, especially the next generation

The authors are greatly indebted to the European Space Agency (ESA) and the National Aeronautics and Space Administration (NASA) who provided many of the photographs for this book. In this regard they are particularly grateful to Willem van der Blas, Elizabeth Laurentin, Martin-Pierre Hubrecht and Simon Vermeer of ESA and Edward Harrison, Debbie Rahn and Richard Underwood of NASA. Other illustrations were provided by: Aeritalia (p.28); M. Ackerman (p.163); Alabama Space & Rocket Center (p.11); B. Bayliss (pp.90, 91); R. Beaujean (p.138); I. Berry (pp.95, 104); J. Bodechtel (p.92); British Aerospace (p.23); Centro Ricerche Fiat (p.147); CNES (pp.170, 176); A. Cogoli (p.150); G. Courtès (p.136); L. Culhane (p.81); DFVLR (p.162); Dornier Systems (pp.158, 159); L. Frank (p.85); A. Gabriel (p.159); M. Gadsden (p.20); J. Harvey (p.81); M. Herse (p.139); Hunting G. & G. (p.90); Inmarsat (p.15); C. Keeling (p.19); M. Mackowski (p.171); Macmillan Journals Ltd (p.70); MBB/ERNO (pp.29, 32, 42, 45, 97, 98, 99, 129, 144, 174, 175); McDonnell Douglas (p.47); L. Napolitano (p.147); National Remote Sensing Centre (p.24); J. Newton (p.78); R. Nitsche/A. Eyer (p.145); G. Newkirk (p.82); Odetics Inc. (p.132); Rockwell International (p.52); Sacramento Peak Observatory (p.82); Thomson – CSF (p.41); Transart (p.25); TRW Systems (p.39); UK Schmidt Telescope Unit, Edinburgh (p.77).

Additionally, the considerable assistance given by Gordon Bolton, Ian Pryke and Marilynne Taylor has been much appreciated.

Finally, they wish to thank all their colleagues, scientists, engineers and other associates on both sides of the Atlantic. Although they are too numerous to be mentioned by name, without their help and support, *Spacelab: research in Earth orbit* would not have materialised.

Published by the Press Syndicate of the University of Cambridge
The Pitt Building, Trumpington Street, Cambridge CB2 1RP
32 East 57th Street, New York, NY 10022, USA
10 Stamford Road, Oakleigh, Melbourne 3166, Australia

First published 1984

Printed in Great Britain by
W.S. Cowell Ltd, Ipswich

Library of Congress catalogue card number: 84-17456

*British Library cataloguing in publication data*
Shapland, David
  Spacelab.
  1.  European Space Agency
  2.  Spacelab Programme
  I.  Title    II. Rycroft, Michael
  629.44′5    TL788.4

ISBN 0 521 26077 9

DS

# CONTENTS

# THE IMPACT OF SPACELAB

The ninth Space Shuttle flight held more than usual significance for Europe in general and the European Space Agency in particular. After a decade of intense development work, a European-built Spacelab, containing, among others, a very varied complement of European experiments, and manned by a team including a European Payload Specialist, has taken its place as a major contribution to the NASA Space Transportation System.

For the European Member States of ESA, this magnificent achievement heralds an exciting future. We can confidently expect that future flights of individual and combined elements of Spacelab will not only consolidate but also enhance the reputation of European design and workmanship.

With thoughts turning to free-flying platforms delivered and retrieved by the Space Shuttle, and even to modular-built Space Stations, Spacelab has given European technologists and industry the possibility of a strong hand in shaping future developments.

It is now up to Europe to capitalise on the firm foundation given by Spacelab.

Erik Quistgaard
Director General, European Space Agency 1980–4

Spacelab is a truly outstanding European achievement, and a symbol of the close cooperation between Europe and the United States in space research and exploration. With the addition of Spacelab, the Space Shuttle now offers the world science community a remarkably versatile laboratory for research in orbit.

Spacelab is indeed revolutionary. With its ingenious modular design, Spacelab permits many combinations of experiments. It allows scientists and engineers to accompany their experiments into space, and enables near-real-time interaction with experimenters on the ground by way of advanced communication and data relay satellites. It makes possible frequent, routine access to space for research with re-use of valuable equipment, and assures recovery of test specimens. I am confident that this new mode of space experimentation, so similar to laboratory research on the ground, will pay off handsomely in ways we cannot yet predict.

James M. Beggs
Administrator, US National Aeronautics and Space Administration

Spacelab is far more than a scientific and engineering triumph of the modern age. It is a symbol of international collaboration in a terrestrial environment that is becoming dominated by military interests. Three decades after World War II, which led to the genesis of the V-2 rocket, the nations of Europe joined to create the European Space Agency, whose mandate is the peaceful exploration of space. This consortium was able to commit itself to an expenditure – about £500 million – that none of the individual nations could have faced alone. The logistic and administrative successes in the further collaboration, leading to the successful integration of Spacelab with the US Space Shuttle, inspires hope for future space enterprises in which global collaboration will be demanded.

Professor Sir Bernard Lovell, FRS

There have been various important 'firsts' during the opening decades of the Space Age. It all began on 4 October 1957, with the flight of Sputnik 1. Then came the first manned journey, by Yuri Gagarin; the first Americans in space; the first space rendezvous; and the pioneer orbital stations, both Soviet and American; to say nothing of the Apollo Program and the surveys of the planets by unmanned probes. The Space Shuttle is of special importance; it may be said to have ushered in the second phase of space exploration.

Now we have Spacelab, which is as important a 'first' as any. It is manned; a fully trained scientist is always present during the mission. Scientific experiments of all kinds can be carried out, all under the control of experts. Moreover, these experts are scientists first and astronauts second. They have the opportunity to work in a laboratory far removed from anything possible on the surface of the Earth.

Spacelab has been tested; it has proved to be an outstanding success, and it augurs well for the future. Therefore, the time is ripe for the first detailed book about it, and this is what David Shapland and Michael Rycroft have produced. It is up to date and authentic; it includes contributions from the first four Spacelab scientists, and it will introduce this all-important mission to readers all over the world. I feel honoured to have been invited to contribute an introductory statement.

Patrick Moore

# FOREWORD

The first Spacelab flight in 1983 represented the fulfilment of a very ambitious programme funded and developed in Europe, in cooperation with the USA.

Following the successful Apollo Mission to the Moon, NASA studied the possibility of the construction of a Space Station in orbit. In 1971–72, these studies were transformed into the project of building a Shuttle to carry payloads into orbit and then return to Earth. NASA suggested to the European Agencies (ESRO and ELDO) the possibility of their being associated with this programme. The outcome of this was Spacelab, a laboratory which would be completely integrated into the Shuttle for flights of up to 10 days. An agreement between ESRO (representing nine of its ten Member States) and NASA was signed in mid-1973 and work started immediately.

Spacelab is one of the most important and most expensive space programmes in Europe. Its total cost is of the order of US$800 million. All European Member States except Sweden decided to participate and were joined later by Austria, an Associate Member State. Relative financial contributions were decided, the main shares being paid by the Federal Republic of Germany (54.9%), Italy (15.6%), France (10.3%) and the United Kingdom (6.5%). Following a tender, VFW–Fokker/ERNO of Bremen was selected to lead the project. An industrial consortium was set up with ten other European companies.

A major problem was the fact that Europe had to develop the Spacelab at the same time as the USA developed the Shuttle. Thus, many specifications for interfaces were 'to be determined'. In some cases, final specifications led to important changes in the design of Spacelab, which needed an additional 2 years and increased funds (40% more than expected) to be completed. Delay in the Shuttle development was, however, even longer, so that Spacelab was ready well before its first flight with the Shuttle. During the coming decade, Spacelab will be taken into orbit three or four times a year, each time equipped with different payloads.

The First Spacelab Payload, half European, half US, contained more than seventy different experiments in various fields. The two fields for which the benefit of Spacelab will be most promising are certainly Life Sciences and Material Sciences.

For Europe, Spacelab is not only a successful achievement, but represents the entry of Europe into manned space activities. The USA has decided to build its first Space Station to be put into orbit in 1992 or 1993. Europe hopes to be associated with this new venture and the experience we have gained in developing Spacelab will be very useful in building certain modules and pallets as part of the Space Station. In this respect it is planned to proceed to new agreements between ESA and NASA.

As ESA Director responsible for Spacelab, I welcome the decision of the Cambridge University Press to publish this book which provides excellent descriptions of the challenge presented by the development of Spacelab and of how Spacelab can be used to further our scientific knowledge.

Michel Bignier
Director of Space Transportation Systems,
European Space Agency

# PREFACES

When I joined the staff of the European Space Agency in 1971, one of my first jobs was to investigate how European industry might contribute to the NASA post-Apollo manned space program. Based on this and on other considerations, a decision was made to design, develop and build the 'Sortie Can' – later to be named Spacelab.

The aim was to make space readily accessible to all experimenters. Early design studies reflected this philosophy by ensuring that laboratory equipment could be used in Spacelab with a minimum of modification. In 1972 I worked with a small, dedicated group from ERNO studying concepts to satisfy the requirements of US and European scientists. The outcome was the modular Spacelab known today.

Many of our early dreams have been dispelled by such things as high launch costs, the high standards of safety needed by a manned system, and the large amount of documentation that must be handled by scientists and engineers alike. A significant high-cost factor arises from the sheer size of Spacelab. It takes over 4 tonnes of experiment equipment to use its capability fully. Although the cost per kilogram of experiment put into orbit is low, the total mission cost is quite high.

Spacelab is a new and exciting way of doing research in orbit. However, costs must be brought down. The great need is for simplification. Everyone concerned must be prepared to take bigger risks because the price of certainty is high. Spacelab is Europe's entry ticket into manned spaceflight which will, no doubt, provide stimulating challenges for the future.

In this book Dr Rycroft and I have tried to present Spacelab and its use in simple language. It is written for the people who paid for it and who will benefit from its use.

D.J.S.
June 1984

Since my student days I have been most interested in Space Science. So, on seeing the advertisement for the position of Payload Specialist aboard the first Spacelab in the back pages of *Nature* in March 1977, I decided to apply in the spirit of 'nothing ventured, nothing gained'. My training, age and height were within the limits specified – though, being 189.5 centimetres tall, only just!

I was interviewed by a distinguished Board, was examined medically and undertook various psychological tests in London in July. The next month, while attending an international scientific conference in the USA, I was very surprised and delighted to receive a telegram informing me that I was one of the five candidates being proposed by the United Kingdom for the Payload Specialist position. After returning home, as one of the fifty-three European candidates, I attended two interview boards at the ESA Headquarters in Paris. Regrettably I did not pass the second.

Nonetheless, I have retained a strong interest in the Spacelab Programme, amongst other European and US space programmes. I was inspired by the first launch of the Space Shuttle to contemplate writing this book. For me it has been a rewarding experience, particularly working with David Shapland whose view of Spacelab is so complementary to mine. Finally, I am deeply grateful to each and every person who, in his or her own valuable way, has helped this book to come to fruition.

M.J.R.
June 1984

## Brief history of spaceflight

Space – the word conjures up many different images in the mind of Man. For some, it is the image of films like *2001. A Space Odyssey* – science fiction. For others, it is the reality of rockets launching satellites into Earth orbit and Man taking his first steps in space – science fact.

Five centuries ago, Ptolemaic astronomy reigned supreme, just as it had for fourteen centuries. The Earth was believed to be stationary; around it revolved the Sun, Moon, planets and stars. But in 1543 came a world-shattering event – the publication of *De Revolutionibus* by Nicolaus Copernicus (1473–1543). The Earth was relegated to being just one of several planets in orbit around the Sun. The invention of the telescope enabled Galileo Galilei (1564–1642) to discover sunspots, to observe the planets and to discover the moons of Jupiter which emphasised the fact that the Earth was not the sole centre of motion in the Universe. The basic laws of planetary motion were formulated by Johannes Kepler (1571–1630), who established that the planets move in elliptical orbits around the Sun. Why this should be so was shown by Sir Isaac Newton (1642–1727) when he constructed his Universal Law of Gravitation. These were the scientific discoveries which laid the foundations for modern astronomy and the development of spaceflight.

# SPACE, OUR LATEST FRONTIER

*'It is difficult to say what is impossible, for the dream of yesterday is the hope of today and the reality of tomorrow.'*
Dr Robert H. Goddard

The planets of the Solar System move in elliptical orbits around the Sun. Our own planet – Earth – is at a distance of approximately 150 million kilometres from the Sun. Mercury, Venus and Mars have rocky surfaces similar to that of the Earth; the giant planets – Jupiter and Saturn – are gaseous; while the outer planets have extremely cold surfaces.

The history of spaceflight itself began with Konstantin Tsiolkovsky (1857–1935), who wrote the first paper on the use of rockets propelled by liquid hydrogen and liquid oxygen. In 1923, Hermann Oberth (b. 1894) wrote a book called *The Rocket into Interplanetary Space (Die Rakete zu den Planetenraümen)* in which some of the problems of propulsion – and of living – in space were discussed. Robert Goddard (1882–1945) independently studied the principles of rocketry, and was the first to construct a liquid-fuel rocket, in 1926. His rockets were propelled by petrol (gasoline) and liquid oxygen. Wernher von Braun (1912–1977) was one of the foremost rocket engineers of recent years. He developed the V-2 rocket and led the team that successfully launched the first US Earth-orbiting satellite.

Robert Goddard in his workshop in New Mexico.

Although much of the pioneering work was carried out in Western Europe, it was the Americans and Soviets who were actually to achieve spaceflight. The world was surprised when, on 4 October 1957, the first man-made object to orbit the Earth was launched by the USSR. This artificial moon was named Sputnik, or 'fellow traveller', following the suggestion of Konstantin Tsiolkovsky. In January 1958, the USA followed suit, with the successful launching of Explorer I. The instruments on this satellite discovered belts of electrically charged atomic particles trapped above the Earth's upper atmosphere by the Earth's magnetic field. These belts of high-energy particles are now known, after their discoverer, as the van Allen radiation belts. The US National Aeronautics and Space Administration (NASA) was also set up in 1958.

On 12 April 1961 the USSR became the first nation to launch a man into space. Yuri Gagarin returned to Earth after completing just one orbit. John F. Kennedy committed the USA to landing a man on the Moon and returning him to Earth, before the end of the decade. As a first step, in February 1962, John Glenn became the first American to orbit the Earth, completing three orbits in his craft, Freedom 7.

The Soviets preferred exploration of the lunar surface by a remotely controlled vehicle, the Lunakhod, to landing a man on the Moon. They used their Soyuz system for taking men to and from their Earth-orbiting station Salyut where they could live and work for extended periods. Meanwhile the Americans progressed through their

William Pickering, James van Allen and Wernher von Braun proudly hold aloft the first US satellite, Explorer 1. They pioneered the use of rockets for scientific research.

Mercury (one-man) and Gemini (two-man) orbital space programmes to the three-man Apollo spacecraft, which were tested in Earth orbit before Apollo 11 set off to land on the Moon in July 1969.

Since the flight of the first Sputnik, more than a thousand satellites have been put into Earth orbit. Scientific research on the near-Earth space environment has progressed apace, notably through the Cosmos, Intercosmos and Explorer series of satellites. The Moon has been investigated through the US Apollo and Soviet Luna programmes. Space probes have now explored Mercury, Venus, Mars, Jupiter and Saturn – planets of our Solar System. Uranus will be visited in January 1986 by the US probe Voyager 2 which will reach Neptune in August 1989.

Thus, Man learned how to use a rocket to overcome the force of gravity and explore space.

This spectacular picture of Saturn and its rings was taken from Voyager 1. Before reaching Saturn, this spacecraft surveyed the other giant planet, Jupiter; it has now left the Solar System.

# Gravity

It is said that one day in 1665 Isaac Newton was sitting under an apple tree when an apple fell on his head. This led him to consider why the Moon does not fall towards the Earth, even though the force governing the motion of the Moon and the fall of the apple is one and the same, the force of gravity. He set about formulating a generally applicable theory of gravitation. There were many difficulties to be overcome, not the least of which was the lack of an adequate mathematical tool for the study. For this Sir Isaac Newton invented the differential calculus. Finally, in 1687, the *Principia* was published, giving his Universal Law of Gravitation. This expresses the magnitude of the gravitational force $F$ between two bodies, of masses $M$ and $m$, whose centres are separated by a distance $r$, as $F = GMm/r^2$. Here, $G$ is called the gravitational constant. Thus, in order to calculate the force of attraction between the Earth and the apple, $M$ becomes the mass of the Earth, $m$ the mass of the apple, and $r$ the radius of the Earth, $R_E$. Now the acceleration of all bodies due to gravity, $g$, at the Earth's surface is 9.8 metres per second per second. This means that the apple will increase its downwards velocity by 9.8 metres per second for every second that it falls. This acceleration is in fact equivalent to a downward force of $mg$. The weight of the apple is $mg$.

If the apple is thrown straight upwards, it will be continuously decelerated by gravity. It will rise more and more slowly, stop, and then fall back again to Earth, all within about a second. An upwards force is exerted in throwing the apple. The greater this force, the higher the apple will rise before falling back to Earth. Thus, in theory, if the apple is thrown up with sufficient force, at some velocity it can escape from the Earth's gravity and disappear into space. This escape velocity, $v_e$, is $\sqrt{(2gR_E)}$, or 11.2 kilometres per second. The escape velocity is independent of the mass, $m$, and is the same for any object, whether it be an atom or a rocket.

In order to put a rocket into orbit, however, a lower velocity would be sufficient. If there were no air resistance, then the most efficient launch would be horizontal, with the rocket moving in a circle of radius $R_E$. Gravity provides the force necessary to keep it in orbit. The orbital velocity is found to be $\sqrt{(gR_E)}$, or 7.9 kilometres per second. However, when air resistance is taken into account, a slightly larger force is required to reach orbital velocities. A rocket must be lifted to a height above 200 kilometres, where the atmosphere is sufficiently rarefied that air resistance causes only slight effects on the rocket's motion.

---

The system of units used here and for most scientific work is the 'Système Internationale' (SI); it is based on the fundamental units of the metre, kilogram and second.

---

Europe's Ariane rocket is launched from Kourou, French Guiana. This rocket is particularly well suited to putting satellites into geostationary orbits.

## Satellite orbits

The orbits of satellites are circular or elliptical. For the Space Shuttle, the orbit is normally circular. For a satellite in an elliptical orbit the highest point is termed apogee, and the lowest point perigee.

The time that a satellite needs to complete one orbit around the Earth, the orbital period, depends on the size of the orbit. For a circular orbit at a height of 300 kilometres, the orbital period is 90 minutes – just an hour and a half! In this time, the Earth moves eastwards through 22.5 degrees of longitude, so that on successive orbits a satellite passes over different parts of the Earth.

An elliptical orbit of a satellite travelling around the Earth. The orbital inclination shown is 57 degrees, the same as that of the first Spacelab flight.

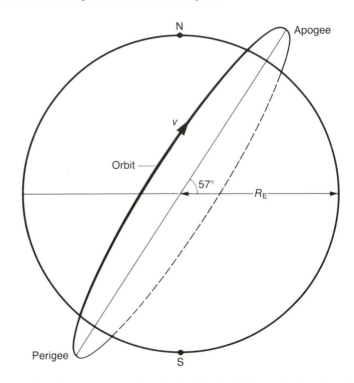

- For a circular orbit of height $h$, the orbital period is

    $$2\pi\,(R_E + h)/v$$

- If the kinetic energy of a rocket due to its motion, $\frac{1}{2}(mv^2)$, exceed the gravitational potential energy at the surface of the Earth, $mgR_E$, the rocket escapes from the Earth's gravitational field. Its velocity must equal or exceed the escape velocity, $v_e$, given by

    $$\frac{1}{2}(mv_e^2) = mgR_E$$

    $$\therefore v_e \quad = \sqrt{(2gR_E)}$$

- The force needed to move a satellite in a circular orbit of radius $R_E$ is $mv^2/R_E$. This is provided by gravity; the force is the weight of the satellite, $mg$

    $$\frac{mv^2}{R_E} = mg$$

    $$\therefore v = \sqrt{(gR_E)}$$

    The resultant force acting on an object inside an orbiting satellite is zero. Thus the object is effectively weightless.

Further away from the Earth, the Earth's gravitational pull on a satellite is less, being inversely proportional to the square of its distance from the centre of the Earth. The orbital velocity is reduced accordingly. For a satellite at a height of 35 800 kilometres the orbital period is 24 hours. Thus, a satellite at a height of about five and a half Earth radii above the Equator, and rotating in the same direction as the Earth, appears to hover in space.

Satellites placed in such a synchronous orbit, termed the geostationary orbit, are most useful. For example, the European Space Agency's Meteosat is located over the Atlantic Ocean for viewing the weather systems which will affect Europe in a day or two. Communications satellites are generally also placed in geostationary orbits. For example, the three Inmarsat satellites provide almost global coverage for ship-to-ship or ship-to-shore radio communications. The coverage is not quite global because ground stations at latitudes greater than 75 degrees North or South cannot 'see' the geostationary satellite. The ultra high frequency (UHF) radio waves used to communicate with the

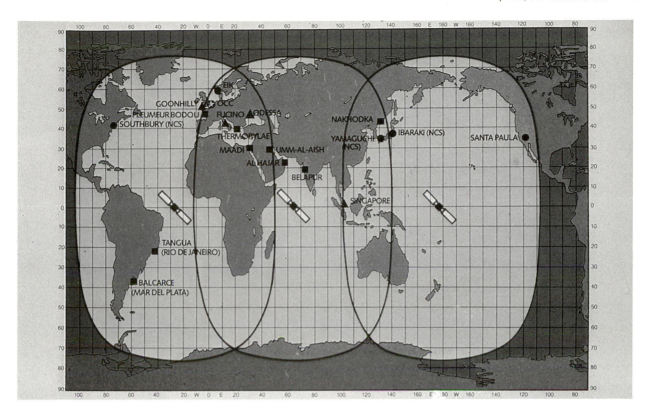

satellites require an uninterrupted line of 'sight'. A small rocket is often used on a geostationary satellite to return it to its correct geostationary position if it drifts off under the influence of solar and lunar gravitational forces. Such a manoeuvre is termed station-keeping.

In general, however, satellite orbits do not lie in the plane of the Equator but can have any inclination to this plane. Rockets are usually launched in the direction of the Earth's rotation, thus gaining an extra 0.4 kilometres per second. To use the Earth's rotation to maximum advantage when putting a satellite into orbit, the orbital inclination should be equal to the value of the geographical latitude of the launch site. Otherwise, extra fuel must be carried by the rocket to change the satellite's orbital inclination.

With an orbital inclination of 90 degrees, a satellite passes over the North and South Poles sixteen times a day for a typical altitude of 300 kilometres. Each time it crosses the Equator, every hour and a half, it does so 22.5 degrees further to the west than on the previous orbit. However, this discussion does not allow for the fact that the Earth – because it rotates – is not exactly spherical. The equatorial radius is about 20 kilometres greater than the polar radius. The Earth's oblateness requires that a satellite at 300 kilometres altitude must have an orbital inclination of 97 degrees (so the highest latitude that it reaches is 83 degrees), in order always to cross the Equator at the same Local Time. This so-called Sun-synchronous orbit is particularly useful when the Earth's surface, or atmosphere, is to be studied from space for some weeks or months under the same conditions of solar illumination. Examples of such remote-sensing studies are agricultural or cartographic investigations, or measurements of the energy radiated into space by the Earth's atmosphere.

The three Inmarsat satellites, in geostationary orbit over the Atlantic, Indian and Pacific Oceans, cover the World's busiest shipping lanes. The different symbols indicate ground stations established at different times.

An image from Meteosat, which is in a geostationary orbit positioned over the Equator on the Greenwich Meridian and observes the weather over Europe and Africa. This image has been artificially coloured.

The thrust of a rocket is measured in newtons in SI units, but it is often quoted in kilograms. To convert kilograms thrust to newtons it is necessary to multiply by $g$ (acceleration due to gravity), the numerical value of which is 9.8 at the Earth's surface.

## Putting a satellite into orbit

Inside a rocket engine, fuel is burnt with oxygen in order to produce a very hot exhaust gas. This gas is ejected downwards at high velocity, producing a force upon the rocket which is in turn accelerated upwards. To launch a rocket, the upward thrust must be greater than the total weight of the vehicle (including fuel) at launch. As it moves upwards, both the vehicle's velocity and its acceleration increase as the burning of the fuel causes its mass to decrease. When all the fuel has been burnt, the rocket will reach a velocity that is directly proportional to the velocity of the exhaust gas. The rocket velocity also depends on the logarithm of the 'mass ratio', defined as the ratio of the initial mass of the rocket, its payload and its fuel, to the final mass of the rocket and payload when all the fuel has been burnt. With a realistic exhaust gas velocity of 2.5 kilometres per second, a mass ratio of 25 is required for the rocket to attain a final velocity of 8 kilometres per second. If gravity and air resistance are allowed for, the mass ratio is nearer 50. In this case only 2% of the total mass of the launch vehicle could be anything other than fuel! It would be impossible to build such a rocket. Even to build a rocket with a mass ratio of 4, where the rocket and payload mass would be 25% of the total mass at launch, requires skilful engineering.

So how can a satellite be put into orbit? The answer involves having a multi-stage rocket. Once the vehicle is moving up through the atmosphere, and when the fuel for the first stage has been expended, the first-stage rocket engine and/or its fuel tanks are ejected and fall back to Earth. Then a second-stage rocket takes over, and accelerates the vehicle which now has a much smaller mass. The mass of the payload is a much larger fraction of the total mass. A third-stage rocket may even be necessary before the payload finally achieves an orbital velocity.

Once in orbit, the satellite is still affected by the tenuous upper atmosphere. Each time it passes near perigee it suffers a retarding force, and so the following apogee is a little lower than the last. The satellite's orbit becomes more circular. However, the braking effect of the atmosphere can be neutralised by using small thrusters on the orbiting vehicle. If this is not done, the satellite will eventually re-enter denser regions of the atmosphere. There, its orbital energy will rapidly be dissipated by friction, and heat will be produced. The outer surface of the satellite will become very hot and might melt and evaporate. The satellite may even break up under the stresses of re-entry. During this dying phase, the deceleration experienced by the satellite may be ten times the acceleration due to gravity. However, any aerodynamic lift during this phase lengthens the descent path. This reduces the deceleration and thus the amount of heat generated at the surface of the satellite, which may then reach the ground without breaking up.

The Earth's polar regions exert a profound influence on the global climate. This is particularly so for Antarctica, a large, high ice-covered continent.

# The Earth's atmosphere

Creatures living on Earth breathe air, of which 21% (by volume) is oxygen. The other major constituent of air is the relatively unreactive gas nitrogen (78%). The inert gases helium, neon, argon and krypton are also present in small quantities. How much water vapour there is in the air is dependent on temperature and locality. Amongst the trace species whose concentrations are measured in parts per million or parts per billion (1 billion = 1 000 million), the most important is carbon dioxide. Taken in by plants and converted into oxygen by photosynthesis, its present concentration is 340 parts per million. However, the concentration of carbon dioxide in the atmosphere is increasing because of the large-scale burning of fossil fuels, such as coal and oil, and also because of deforestation, particularly in the tropics. From 260 parts per million in 1850, the concentration is expected to reach 500 parts per million in the middle of the twenty-first century. This increase could, by the infra-red 'greenhouse effect', cause atmospheric temperatures to rise by 2 degrees Celsius (Centigrade) or more. Such a warming could produce at least a partial melting of the polar ice-caps, and lead to catastrophic changes in sea level and in the global climate.

Atmospheric atoms and molecules are retained by the Earth's gravitational field. The air pressure at ground level is equal to the force exerted by the weight of all the air in a column above a unit area. Air pressure at ground level is thus about one kilogram per square centimetre, known as a bar, or 1000 millibars. The pressure of air is also proportional to the number of molecules per unit volume, that is the number density, and to the temperature, on the Kelvin scale. This 'absolute temperature' is measured in degrees above 'absolute zero'. It is obtained by adding 273 to the temperature in degrees Celsius.

Higher up in the atmosphere, the pressure and the number density of molecules decrease. There is less air above a certain level, the higher that level is. At a height of about 5 kilometres, the atmospheric pressure is half its sea-level value; at twice this height the pressure is halved again, and so on; in other words, the pressure decreases exponentially as the height increases. About three-quarters of the Earth's atmosphere is contained in the lowest 10 kilometres. This is called the troposphere. Weather is generated in this region. About 99% of the Earth's atmosphere lies below the 30-kilometre level. The atmospheric 'skin' of planet Earth has a thickness that is much less than the radius of the Earth.

At levels in the atmosphere where solar energy is absorbed, the air temperature is relatively high. Most solar energy is absorbed by the ground, warming the adjacent layers of air. When a volume of air rises due to buoyancy, it expands. Its pressure thereby decreases, and so does its temperature. The temperature decrease is between 5 and 10 degrees Celsius per kilometre. This is why air at the tops of mountains is colder than air at sea level. At a height of one kilometre, it is often cold enough for the water vapour in the atmosphere to condense and clouds are produced. The water droplets so formed may grow by collision with other droplets, eventually producing rain. Snow occurs where the temperature of the cloud – and the air at ground level too – are below zero degrees Celsius, the freezing point of water. The decrease of temperature with height results in the troposphere being a region of vertical instability. It is a turbulent and windy region.

The upper boundary of the troposphere is the tropopause, above which is the stratosphere. At heights between 20 and 50 kilometres, oxygen molecules strongly absorb ultra-violet radiation from the Sun. Some split into atoms which attach themselves to other oxygen molecules, forming ozone. The ozone also absorbs in the ultra-violet region of the spectrum, causing further heating. Although only one in a million air molecules is of ozone, it is this ozone-rich layer which prevents harmful ultra-violet radiation from reaching the fauna and flora living on the Earth's surface. The temperature increases as height increases through this ozone-rich region, counteracting the effect of buoyancy. It is therefore a stable region. For the safety and comfort of passengers, airliners nowadays fly at a height of 12 kilometres, just above the tropopause. Only a very violent volcanic eruption, such as that of the Mexican El Chichón volcano from the end of March to early April 1982, has sufficient energy to thrust dust particles right up into the stratosphere.

Above the stratosphere is the mesosphere which, like the troposphere, is an unstable region where the temperature decreases with increasing height. By far the coldest region of the atmosphere is found at the top of the mesosphere, and is called the mesopause. Surprisingly the mesopause at high latitudes is coldest in summer, allowing thin cloud layers to be produced. These 'noctilucent' clouds are seen when illuminated by the Sun at low angles of elevation.

Higher yet is the thermosphere where the most energetic radiation from the Sun, that is X-ray radiation, is absorbed by the tenuous uppermost atmosphere – at a height of about 90 kilometres the density of air is only a millionth of its value at sea level. This energetic radiation causes an electrically charged gas to be produced. Formed of electrons and positive ions, this charged gas is termed a plasma. This ionospheric plasma strongly affects the propagation of radio waves. The ionosphere merges with the magnetosphere above it. There the Earth's magnetic field plays a crucial role in determining the motion of the electrically charged gas and of the van Allen radiation belt particles. In contrast, the ultra-rarefied neutral gas in the exosphere is not affected by the magnetic field.

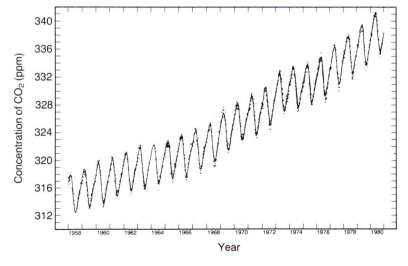

The concentration of carbon dioxide (in parts per million) in the atmosphere, as observed in Hawaii, is continually increasing. The annual variation shown is ascribed to the photosynthesis of plants growing on the island.

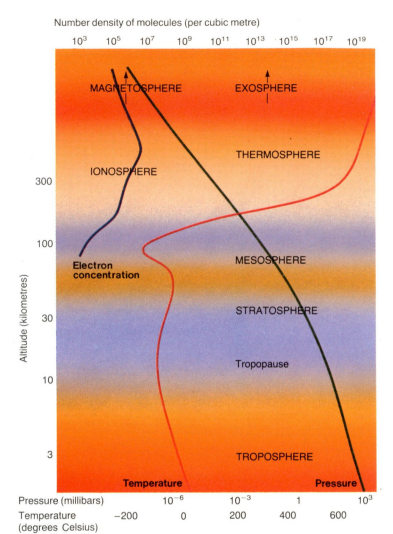

The atmospheric pressure and number density of molecules, shown in green, decrease smoothly with increasing altitude. The concentration of electrons, shown in blue, and the temperature, shown in red, also vary with altitude, but in more complicated ways. The relatively cold regions are shown in blue, warm in orange, and hot in red.

Noctilucent clouds, probably ice crystals, photographed above Aberdeen. They are produced at the summer mesopause, at an altitude of 85 kilometres. The temperature there is about 180 Kelvin, the mesopause being the coldest part of the atmosphere.

# The electromagnetic spectrum

Much of the electromagnetic radiation from the Sun and stars is prevented from ever reaching ground level by the highest layers of the atmosphere. There are only two atmospheric 'windows' which allow radiation to pass through. One is the visible part of the spectrum, light to which human eyes are sensitive. This window extends into the near infra-red. The other window is in the radio part of the spectrum, at wavelengths from a million to a billion times longer than the wavelengths of yellow light.

Radiation can also be specified in terms of its frequency. In free space, the wavelength ($\lambda$) of electromagnetic radiation is related to its frequency ($f$) by the relation $f = c/\lambda$. Here, $c$ is the velocity of light, or of any other electromagnetic radiation, in vacuo, and is about $3 \times 10^8$ metres per second. Thus the frequency of yellow light, of wavelength $6 \times 10^{-7}$ metres, is $5 \times 10^{14}$ cycles per second, or hertz. The limits of the radio window are one million hertz (one megahertz) and one billion hertz (one gigahertz). The radio window thus extends from high frequency (HF) through very high frequency (VHF) up to ultra high frequency (UHF).

For this reason, both FM (frequency-modulated) radio and television transmissions from Earth, often at frequencies near 100 megahertz, continue out into space, where they could be picked up by other civilisations who had sufficiently sensitive radio equipment. Only radio telescopes and optical telescopes can observe the Universe from the surface of the Earth. Observations at all other wavelengths need instruments on rockets or satellites viewing the Cosmos from above the obscuring atmosphere.

## Reporting numbers great and small

| | index notation | prefix |
|---|---|---|
| one thousand (1000) | $10^3$ | kilo |
| one million (1 000 000) | $10^6$ | mega |
| one billion (1 000 000 000) | $10^9$ | giga |
| one-thousandth (1/1000) | $10^{-3}$ | milli |
| one-millionth (1/1000 000) | $10^{-6}$ | micro |
| one-billionth (1/1000 000 000) | $10^{-9}$ | nano |

Radiation from the Sun and the Universe reaches the Earth's surface only if it is in certain wavelength bands. The atmosphere has two main 'windows', in the visible and radio regions of the spectrum. All other radiation (of different frequencies – and wavelengths) is, to a greater or lesser extent, absorbed or reflected by different regions of the Earth's atmosphere.

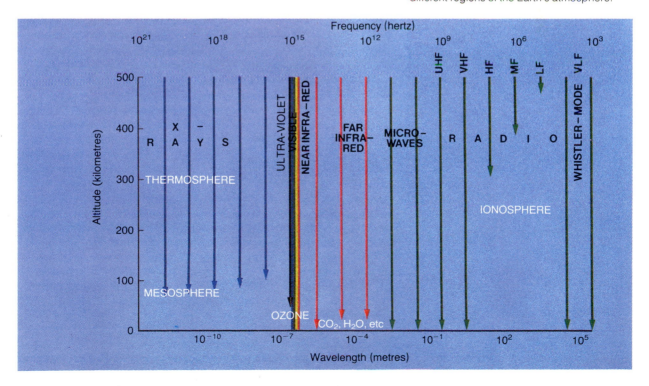

# Benefits of space to Mankind

Human curiosity and the desire to explore provide the basic motivations for space research. Man searches to place his species in context within the Solar System and the Universe. Man seeks to discover, and to understand as far as possible, the physical laws which govern the behaviour of matter in the Universe. Above the obscuring and distorting atmosphere, his instruments can explore the vastness of the Universe at all wavelengths of the electromagnetic spectrum.

The part of the electromagnetic spectrum in which objects radiate most energy is related to their temperature. Objects at one million Kelvin, for example, radiate predominantly in the X-ray region, at wavelengths around 3 nanometres. The Sun's surface, whose temperature is 6000 Kelvin, radiates in the visible region. Relatively cool interstellar clouds of gas and dust, where stars are continually being formed, shine brightly in the infra-red at 30 micrometres. The temperature in such clouds is about 100 Kelvin. Radiation at 3 Kelvin results from the sudden creation of the Universe about 20 billion years ago. Present ideas of this 'Big Bang' suggest that the Universe was once very small and had a temperature of $10^{13}$ Kelvin. Ever since then, the Universe has continued to expand and to cool.

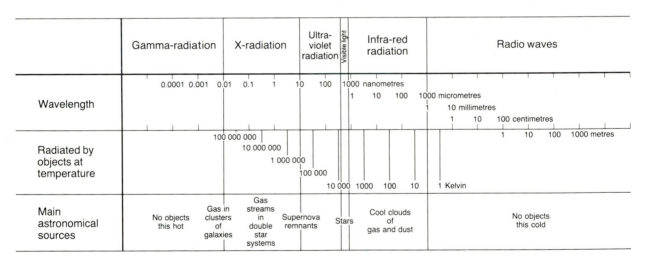

Various objects in the Universe are at different temperatures. The higher the temperature, the higher is the frequency (or the shorter the wavelength) of the emitted radiation.

Man needs to understand the workings of planet Earth, its oceans and its atmosphere. Such understanding is necessary on a broad scale before any improvements can be made to his interpretation of environmental observations and to his subsequent forecasting. Accurate weather forecasts would be much appreciated by everyone and would help not only in the arranging of day-to-day activities but also in matters of global concern such as the planning of agriculture. Another concern is the effects of pollution on the Earth's environment. Substances introduced into the atmosphere by Man could have a profound influence on world climate. Changes of climate would affect the daily lives of all. In particular, the ozone layer must not be harmed. The influences on stratospheric ozone of high-flying aircraft and of the freons used as propellants in aerosol spray cans must be understood. The role of solar activity, particularly that of solar flares, in all these subjects must also be established.

Forecasts will be required of geomagnetic storms, only some of which follow solar flares, but all of which are associated with brilliant auroral displays. These spectacular lights are seen in the night sky at heights just greater than 100 kilometres. Occurring at high latitudes in the northern and southern hemispheres, they are called the Aurora Borealis and the Aurora Australis, respectively. The storm-related aurorae occur at lower latitudes than usual. During aurorae high voltages are induced in large-scale conductors such as power lines for distributing electricity, trans-oceanic cables, telephone lines and steel pipelines. These high voltages cause damage to power transformers; they also cause circuit breakers to trip and hence produce power cuts. Also, they disrupt radio and telephone communications, and result in excessive corrosion of pipelines. Geophysical exploration for hydrocarbon and mineral resources using sensitive magnetic instruments is interrupted during such storms.

Satellites orbiting the Earth can make measurements in the only natural plasma laboratory available to Man to test and develop theories of plasma physics. These theories are applicable throughout the Universe, since 99% of matter in the Universe is in the plasma state rather than in the solid, liquid and gaseous states familiar on Earth. The theories are also relevant to laboratory plasma devices. Some of these may, in the future, generate cheap electricity by the thermonuclear fusion of hydrogen, which is available in plenty in the form of water.

Satellites themselves are affected when they pass through the belts of energetic charged particles trapped around the Earth. In particular, their microelectronic circuits can suffer from radiation damage. They also need to be protected from sudden electrical charging. Geostationary satellites are particularly prone to this potentially damaging event.

Examples of geostationary satellites are satellites for intercontinental radio communication (voice, telex and data) at frequencies measured in gigahertz, for direct broadcasting of television to the population of an entire continent, and for observing the evolution of weather systems. For the future, space power stations would be set up in a geostationary orbit. These might consist of solar cells in an array of some 50 square kilometres and costing several billion dollars. The electrical energy thus produced would be converted into microwave energy and beamed to Earth. The beam, with a typical diameter of about 7 kilometres, would have an energy flux of 100 watts per square metre, that is about 7% of the solar flux at the Earth's surface. However, concern has been expressed about the heating effects that this microwave flux would have on the thermosphere and ionosphere.

The European Communications Satellite (ECS) was aunched by Ariane into a geostationary orbit in June 1983.

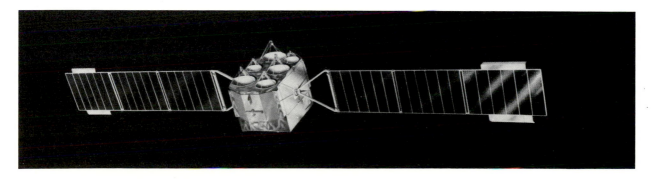

The mysterious aurora, here observed in northern Sweden, is produced at altitudes near 110 kilometres by energetic electrons bombarding the uppermost atmosphere.

Satellites can be used for 'remote sensing' of the Earth's surface, both land and sea, in the visible or infra-red parts of the spectrum. Remote sensing has applications in agriculture, cartography, fishing, forestry, geology, geophysics, hydrology, meteorology, oceanography and for studies of pollution. Also satellites can carry radars of several types and degrees of sophistication for many geophysical studies. Several polar-orbiting satellites can be used as a navigational system, providing position fixes for ships and aircraft.

Life sciences and material sciences experiments can be performed in space under weightless conditions. For example, experiments can be carried out in metallurgy and into the behaviour of liquids and biological materials. Pure drugs, enzymes or hormones, or large near-perfect crystals of use in the semiconductor industry could be manufactured. Man's presence aboard a space vehicle is most valuable in such fields. Some experiments can also be performed on Man himself.

Man's innate curiosity, aroused so long ago, has so far been only partially satisfied by space exploration. His intellect is being challenged by every new venture. From the exploration of space Man also derives many tangible benefits. High-precision equipment is continuously being developed for spaceflight. The technological advances involved are soon brought into equipment manufactured for industry and consumers on Earth.

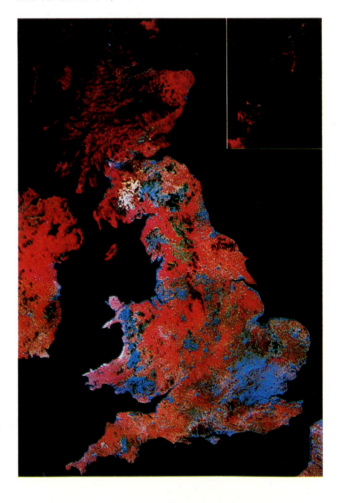

Satellite observations made in different regions of the spectrum are processed to reveal many features of the Earth's surface. In this artificially coloured composite picture of the British Isles.

## The Spacelab system

Spacelab is a laboratory in which scientists can perform experiments in space. It allows them to use ordinary laboratory equipment in space in a manner that is not possible with conventional automated satellites. Spacelab has therefore been designed to provide the essentials for doing this. These are:

(i) the capability to transport and house an adequate mass and volume of equipment

(ii) the provision of power and the thermal control for that equipment

(iii) the means of sending command signals to the experiments, and of obtaining data from the experiments

(iv) the supply of suitable specialised equipment, such as airlocks and pointing systems for instruments

(v) the provision of a computer system and associated software for managing the supporting subsystems, the experiments and, particularly, the flow of data

(vi) the presence of Man to oversee the operation of the subsystems and experiments, to intervene in the experiments if necessary, and to repair faulty equipment.

These provisions are presented to the experimenter in a standard way, so that he can use them much as he does in his own laboratory.

Spacelab is a large and well-equipped laboratory – much larger than the Apollo and Mercury capsules – and much more sophisticated than the Skylab of 10 years ago. Whereas Skylab used existing space hardware, Spacelab is the first purpose-built space laboratory.

# SPACELAB – MADE IN EUROPE

*'Spacelab, Europe's contribution to the Space Shuttle Program, is an extraordinary achievement. It's also the largest cooperative space project ever undertaken.'*

Vice-President George Bush

Spacelab's modular construction is based on two essential elements – the module and the pallet. The sizes of both module and pallet are adjusted to suit a particular mission. The module is cylindrical, of diameter 4 metres, and consists of either one or two 2.7-metre-long segments enclosed between end cones. The configuration is called a 'short module' or a 'long module' depending on the number of segments used. The overall lengths are about 4.3 metres and 7 metres, respectively. The module fits inside the cargo bay of the Space Shuttle. It is fixed at the bottom of the bay and at the sides.

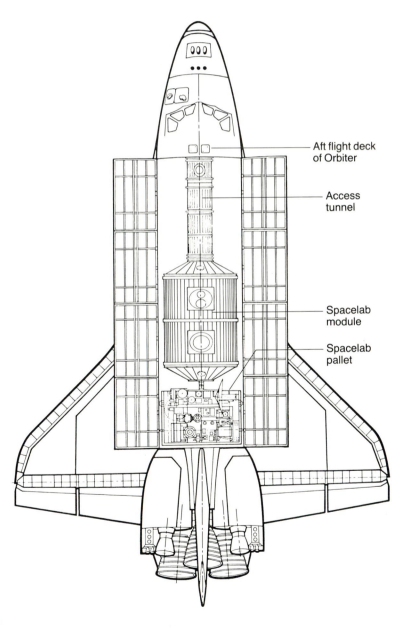

Aft flight deck
of Orbiter

Access
tunnel

Spacelab
module

Spacelab
pallet

Drawn to scale, this diagram shows how the module and pallet of Spacelab are tailored to fit the Orbiter's cargo bay. The bay is 18.3 metres long and 4.6 metres in diameter.

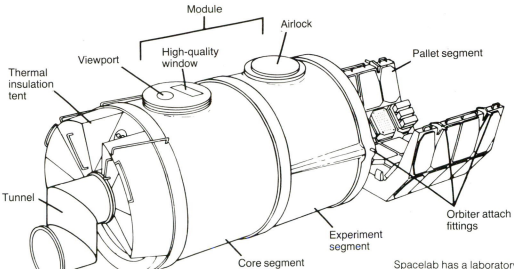

The single segment in the 'short module' is the 'core segment'. This contains the Spacelab subsystems which provide the crew with all the necessary laboratory facilities. Eight cubic metres of the core segment are available for equipment for the experiments. If more space is needed for the experiments, a second segment is added to the core segment. This 'experiment segment' is dedicated solely to experiments and provides a further 14.6 cubic metres for housing equipment. In the module, experimental equipment is housed in standard racks that take the usual 19-inch (48-centimetre) size laboratory equipment trays, or it is mounted centrally on the floor (centre aisle). The fixtures and connections for equipment are standard as far as possible; a Payload Accommodation Handbook has been produced to help experimenters to interface their equipment with Spacelab.

The crew reach the module from the Orbiter cabin through a tunnel one metre in diameter. Both the tunnel and the module are kept at a comfortable temperature and humidity with a normal oxygen–nitrogen atmosphere. The crew can thus work 'in their shirt sleeves' under these Earth-like laboratory conditions.

A limited amount of Spacelab-related work can be carried out in the aft flight deck of the Orbiter. Computer terminals are available there for the scientist–astronauts so that they can watch displays duplicating those in Spacelab itself. The Spacelab crew can talk to each other, with the Orbiter crew, and with the ground-based experimenters, using a specially designed 'cordless' intercom. Closed-circuit television enables the crew to survey the Spacelab equipment.

The pallet is also modular, being built up of segments 3 metres long and 4 metres wide. These may be mounted individually or up to three may be joined together to form a single 'pallet train'. The pallet cross-section is U-shaped.

Spacelab is extremely versatile because of the modular approach adopted. Variations upon the three basic flight configurations – module only, module and pallet, and pallet only – can be obtained by changing the number of module segments or pallet segments.

Spacelab has a laboratory part (the module) in which scientists can work in 'shirt-sleeve' comfort and an observatory part (the pallet) where experiments can be exposed directly to the high vacuum of space.

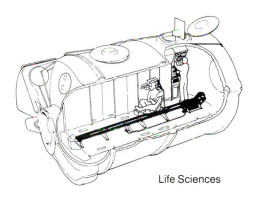

Life Sciences

A module-only configuration (above) is particularly suitable for Life Sciences experiments. The pallet-only mode (below) is more appropriate for Astronomy experiments.

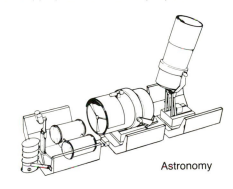

Astronomy

The module of Spacelab, covered by its heat-insulating blanket, is shown here with two pallets.

The outer shell of the module is cylindrical and made of aluminium alloy. The holes that will eventually take the window and airlock are at the top.

In the pallet-only mode, the essential subsystems such as computers and data-handling equipment (which are normally part of the core segment) are contained in an 'igloo', ensuring a pressurised, temperature-controlled environment. The igloo is a cylinder 1.1 metres in diameter mounted to the front frame of the first pallet. Experimental equipment mounted on the pallet can be controlled from the module, from the aft flight deck of the Orbiter, or from the ground. Pallet-mounted experiments can also be adjusted by spacesuited members of the crew during 'spacewalks'.

The centre of gravity of the Spacelab–Orbiter combination must be strictly controlled so that safety is ensured when re-entering, flying within the atmosphere, and landing. Spacelab must therefore be placed towards the back of the cargo bay of the Orbiter.

The total mass of Spacelab and its payload is limited to 14 500 kilograms for landing. However, additional mass, such as equipment to be left in orbit, or consumables, may be carried at lift-off. Although the mass of the system at launch depends on the exact configuration, the following values for some Spacelab elements give an idea of their contributions to the total mass (in kilograms):

| | |
|---|---|
| core module segment (structure) | 685 |
| aft/forward end cone (structure) | 255/295 |
| total short module (with floors, etc.) | 1755 |
| pallet segment (structure) | 590 |
| core module subsystems | 2000 |
| single/double rack | 42/59 |
| pallet segment subsystems | 135 |
| tunnel | 880 |
| crew allowance (one Payload Specialist + equipment) | 215 |
| Instrument Pointing System | 1265 |
| igloo (equipped) | 640 |

In the pallet-only mode of using Spacelab there is no module for housing the main subsystems. In this case, they are enclosed in a pressurised cylinder called an 'igloo'. The igloo, attached to a pallet, is shown here being tested before the lid is put on.

During the assembly of the various Spacelab parts in ERNO's Integration Hall, large items could only be moved using cranes. Here a core module is being taken from one part of the assembly area to another.

(Below) When instruments need to be exposed to space, they are mounted on a pallet element. The Orbiter's cargo bay can house up to five of these 3-metre-long elements.

## Spacelab resources available for experiments

| | Configurations using module | Pallet-only configurations |
|---|---|---|
| Payload mass (kilograms) | up to 4900 (long module) | up to 8000 |
| Volume for experimental equipment (cubic metres) | | |
|     Pressurised | 7.6 (short), 22.2 (long) | – |
|     Non-pressurised | ←———————— about 33.5 per pallet ————————→ | |
| Pallet mounting area (square metres) | ←——————— about 17 per pallet ———————→ | |
| Electrical power (28 volts DC, 115/200 volts at 400 hertz AC) | | |
|     Average (kilowatts) | 3–4 | 4–5 |
|     Peak (kilowatts) | 8–9 | 10 |
|     Total energy (kilowatt hours) | about 400 | about 600 |
| Thermal control (kilowatts) | | |
|     Module atmosphere | 2.7 | – |
|     Cooling for rack-mounted equipment | 4.5 | – |
|     Experiment heat exchanger | 4 | 4 |
|     Cold plate (each 50 centimetres × 40 centimetres) | 1 | 1–8 |
| Experiment Support Computer | ←——— 64 000 (64 k) core memory of 16 bit words ———→ | |
| With Central Processing Unit | ←——— 320 000 operations per second ———→ | |
| Data handling | | |
|     Real time through Orbiter (Ku band) | ←——— up to 50 million bits per second ———→ | |
|     Storage in High Data Rate Recorder | ←——— up to 32 million bits per second ———→ | |
|     Total capacity | ←——— 38 thousand million bits ———→ | |
| Data display | | |
|     Data Display Unit | 999-symbol, 3-colour display on 12-inch diagonal screen | Similar units available in the aft flight deck of the Orbiter |
|     Keyboard | alpha-numeric keys, plus 25 function keys | |
| Instrument Pointing System (mounted on pallet) | ←——— arc second pointing accuracy ———→ in three axes, for up to 3000 kilogram, 3-metre-diameter payloads | |

# Spacelab structure

The module is constructed of a special aluminium alloy and is pressurised, like a modern airliner, to withstand a pressure of one atmosphere (one bar). Thus the crew may breathe the same 'air' as they are accustomed to at ground level.

 The module is designed with ground and space activities in mind. Under weightless conditions a sense of direction within the working area is provided by a 'floor', even though there is no 'up' or 'down' in space. Foot restraints and handrails throughout the module help the crewmen to get around. For 'spacewalks', there are handrails to help the astronaut on the outside of the module and on the pallet.

 Inside the module, the experiment racks are designed to accept most types of laboratory equipment. They are in two forms, 'single' and 'double', of widths 56 and 105 centimetres, and 76 centimetres deep. Besides being attached to the floor, they are also attached to the 'roof', being shaped at the top to fit the contours of Spacelab. The single and double racks can support 290 and 580 kilograms of equipment, respectively. Each rack has its own air ducts for cooling, and electrical power lines and data lines. A full load of racks, either for a core segment or for an experiment segment, consists of two double racks and a single rack on each side. A long module thus holds twelve racks. The first double rack on each side of the core segment contains the Spacelab subsystems (mainly electronics) and a work bench for the crew.

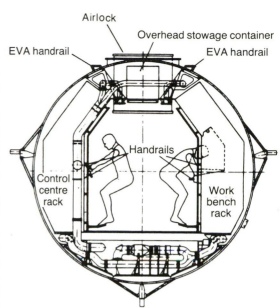

Spacelab's module is designed to give a good working environment for both crew and equipment.

Spacelab's racks are designed to take equipment mounted in standard 19-inch-wide (48 centimetres) trays. Here the racks are being prepared for sliding into the module shell.

The work bench (left) and control rack (right) are shown in this view of the core module being integrated. Noisy equipment, such as fans, is placed beneath the module floor.

Technicians adjust air flow rates for the cooling of electronic equipment placed in Spacelab's racks. Air is blown over the warm equipment and sucked back via the vertical return ducts. This air cooling of equipment is similar to the method employed in an Earth-bound laboratory.

The work bench has storage facilities, a filing cabinet, writing equipment and waste disposal units. Tools such as spanners, screwdrivers and pliers are included for repairing subsystems, instruments or even satellites. Straps and bungees (elastic cords) are distributed throughout the module – particularly at the work bench – for the restraint of articles under weightless conditions.

Lighting can be controlled by the crew throughout the module; lighting levels can be reduced progressively to zero for handling sensitive photographic film or reading faint displays. Emergency lighting is available in case there is a loss of main power.

The first double rack of the core segment also contains a control centre, for monitoring and controlling of the Spacelab subsystems and experiments via a keyboard, and a visual Data Display Unit. The control centre rack and its displays have been designed so that the most important and frequently used controls are placed in the most convenient positions. For this, information on the posture that man naturally assumes under zero-g conditions was evaluated from water-tank trials and from photographs taken during the Skylab missions.

Special cooling equipment may be installed on the double rack next to the control centre rack; this may be a cold, water-cooled plate, or a conventional heat exchanger which also uses a water loop to carry away the heat.

The core segment has two windows, one of which is of high-quality optical glass. There is also a vent for dumping unwanted gases from experiments into space. Films may be stored in special containers, placed in foam inserts, with humidifying chemicals. These containers may be used to protect film from radiation such as that from the van Allen radiation belts. Very noisy items, such as fans and pumps, are fitted underneath the floor.

(Above) The airlock, used for putting experiments into space and then bringing them back into the module, is swung into position for installation in the roof of the experiment module. The airlock is one metre long and has a diameter of one metre. Power, lighting and data-handling connections are at hand.

(Left) The spacious interior of the Spacelab module, ready for flight, shows the control centre rack with its Data Display Unit (on the left), the front panels of rack-mounted equipment and the airlock in the roof. The yellow handrails help the astronauts both to move around the module and to steady themselves in near-weightless conditions.

An airlock can be fitted to the roof of the experiment segment. Using this, an instrument of mass up to 100 kilograms may be attached to a small table, which can be raised, and the instrument exposed directly to space. It may then be retracted and, after the airlock has been repressurised with nitrogen, be brought back inside the module.

Safety is an important factor in any manned space system. A caution and warning subsystem, in the form of a master alarm and display of the relevant data to the crew of both the Orbiter and Spacelab, provides an indication of potentially dangerous situations. Fire detectors, portable fire extinguishers and portable oxygen systems are provided for emergency use.

The U-shaped pallet segments are of conventional aircraft construction and can support 3000 kilograms of experiment equipment for direct exposure to the space environment. The basic frame is covered by honeycomb sandwich panels, of which there are twenty-four per pallet segment. Each square metre of panel can support small instruments of up to 50 kilograms, whereas larger equipment must be attached to the pallet 'hard points', of which there can be twenty-four per segment.

Each pallet segment provides electrical power and signal lines. Freon liquid cooling is available through standard cold plates. Hence, an experiment can operate on the pallet just as it would in vacuo in a laboratory. The crew member operating equipment on the pallet does so through the keyboard of the control centre in the module, or through the keyboard on the aft flight deck of the Orbiter.

# Controlling the environment within Spacelab

A comfortable environment like that of a pleasant summer's day is provided for up to four crew members by the Spacelab Environmental Control Subsystem (ECS). An atmosphere of oxygen and nitrogen is maintained at a temperature of about 22 degrees Celsius (72 degrees Fahrenheit), and a relative humidity of about 50%, by a condensing heat exchanger. Excess water is separated from the air centrifugally and stored. Gaseous oxygen is obtained from the Orbiter, and gaseous nitrogen from a gas tank which is part of Spacelab. Regulators control the pressure and the relative proportions of the gases. Lithium hydroxide, with a small amount of activated charcoal, is used to control the concentration of carbon dioxide and to remove odours and traces of contaminants. Air circulation so necessary for comfort is produced by a system of fans maintaining a filtered air stream moving at speeds of between 5 and 12 metres per minute.

Keeping the crew comfortable is not the only purpose of the ECS. Most of the experiments and components aboard Spacelab will only work within a certain range of temperatures. Thermal control by the ECS is particularly necessary for packages exposed to deep space, which will cool down; other packages, exposed to the Sun, will be heated. There is no difficulty in providing heat where it is needed. Cooling, on the other hand, is a major function of the ECS. The coolants used to carry excess heat to the Orbiter are air, water and freon; excess heat is radiated into space by the large radiators inside the cargo bay doors.

This diagram of the Spacelab Environmental Control Subsystem indicates the four methods of cooling equipment in the module or on the pallet. It is the cabin loop which maintains the comfortable working environment within Spacelab.

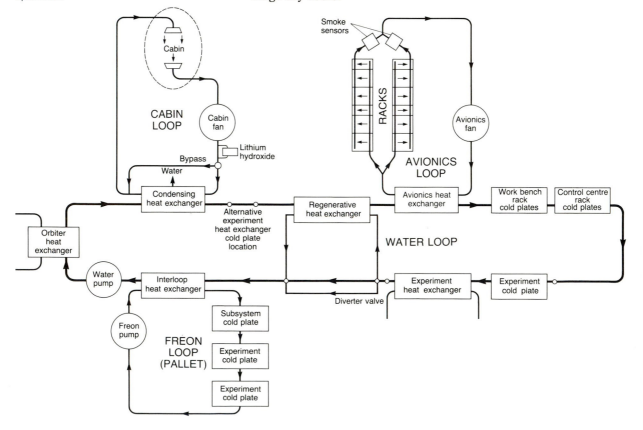

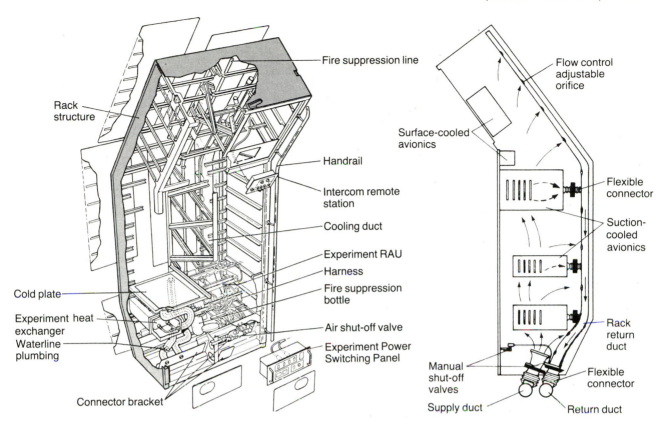

Labels (left diagram): Fire suppression line, Rack structure, Handrail, Intercom remote station, Cooling duct, Experiment RAU, Harness, Fire suppression bottle, Air shut-off valve, Experiment Power Switching Panel, Cold plate, Experiment heat exchanger, Waterline plumbing, Connector bracket

Labels (right diagram): Flow control adjustable orifice, Surface-cooled avionics, Flexible connector, Suction-cooled avionics, Manual shut-off valves, Rack return duct, Flexible connector, Supply duct, Return duct

The equipment in each rack is cooled by a stream of air, as shown diagrammatically on the right. A special rack in the core segment, illustrated on the left, can be fitted with a cold plate or an experiment heat exchanger to deal with experiments that produce large quantities of heat.

Air is used to cool the interior of the module, since heat is generated by the crew themselves and by some subsystems and experiments. Air is also used to cool experiments and electronic equipment mounted in the racks. The air which has been warmed by the operating equipment is sucked into a heat exchanger where the heat is transferred to the cooling water. Valves, normally set before the flight, control the various airflows past the equipment. Laboratory equipment is usually air-cooled and so needs little or no modification for use in Spacelab. 'Direct contact' cooling can also be provided in the module by a heat exchanger where the water directly cools the equipment, or by mounting the equipment on a water-cooled 'cold plate'. Freon, a commonly used refrigerant, is not allowed in the module because of the possible danger to the crew.

On the pallet, air cooling is impossible, and water would freeze, so freon-cooled cold plates are used. The freon cooling system can cope with up to eight standard cold plates. A freon heat exchanger can also be used, in special circumstances, but this is not standard and it would need additional plumbing. Also thermal capacitors can store large quantities of heat by melting wax.

The total heat rejection capability of Spacelab is 8.5 kilowatts continuously, but 12.4 kilowatts for 15 minutes every 3 hours. Spacelab and its payload, as a system, radiate heat into deep space and must be protected from the extreme cold. The system must also be prevented from overheating when warmed by the Sun and, to a lesser extent, by the Earth. Insulating blankets and surface coatings are used to control the temperature of the various pieces of equipment by passive thermal control.

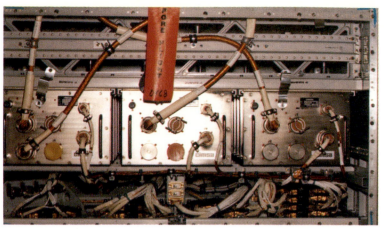

(Above) Each Spacelab rack is equipped with an Experiment Power Switching Panel, shown in the lower right-hand corner of this double rack.

(Above right) The complexity of distributing Spacelab's electrical power is illustrated by this view of the back of the control centre rack. The three CIMSA computers are in the centre.

(Right) Subsystem equipment beneath the floor of the core module is mainly associated with the Environmental Control Subsystem. The experiment racks and floor are in position to be rolled into the shell of the core module. This 'roll on – roll off' feature permits easy access to the experiments and ensures simple loading and unloading of the racks.

# Electrical power distribution

Most experiments require electricity. Spacelab provides its electrical power throughout the laboratory just as in a laboratory on Earth. The fuel cells in the Orbiter are used to generate electricity through the combination of oxygen and hydrogen. The power generated per cell is 7 kilowatts of direct current (DC) at around 28 volts. Almost twice as much power may be extracted for periods of up to 15 minutes. The power from the Orbiter is both stabilised and converted into alternating current (AC) aboard Spacelab so that both DC and AC are available. As is usual on aircraft, an AC supply at 400 hertz is provided by an inverter at 115 volts (single-phase) and 200 volts (three-phase).

An electrical harness is used to distribute this power to all experiment and subsystem equipment, whether in racks or mounted on the pallet. The experiments are supplied via computer-controlled Experiment Power Distribution Boxes (EPDBs). There is an EPDB for each pallet, one in the core segment and two in the experiment segment. An Experiment Power Switching Panel is located in each rack; the required type of power is selected manually using switches located on the front panel of the rack. On the pallet, the experiments interface directly with the EPDBs.

Electrical power is also supplied to experiments located in the centre aisle and airlock, and to equipment attached to the high-quality optical window. In addition, a 750-watt power supply is available for Spacelab purposes in the aft flight deck of the Orbiter. Interior lighting of the module and other low-power requirements, such as caution and warning signals, are also provided. In case of emergency, there is a separate 400-watt supply to essential subsystems and experiments. When the Orbiter is taking off or landing, the total power available in Spacelab is only one kilowatt; most subsystems and experiments are therefore switched off at these times.

The Orbiter's fuel cells provide basic electrical power Spacelab's Electrical Power Distribution System changes this into forms which individual experiments can use.

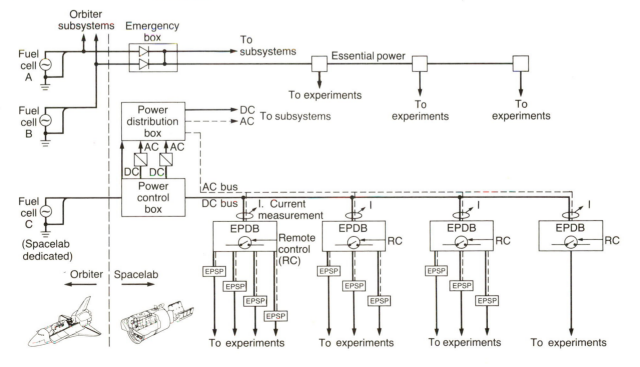

The total amount of electrical energy that can be generated depends on the quantities of hydrogen and oxygen available, as an 'energy kit', for the fuel cell. Normally, 840 kilowatt hours are provided for a mission by one energy kit. Additional energy can be provided by adding further energy kits, but only at the expense of reducing the payload by 740 kilograms per kit.

## Obtaining data from Spacelab

The information generated by experiments on board Spacelab is naturally wanted by the experimenter. Scientists and engineers likewise require certain 'house-keeping' data from Spacelab. Such information may be transmitted to the ground without any significant time delay, that is in 'real time', or stored on board and provided to the experimenter after landing.

Real-time data are transferred using NASA's Space Tracking and Data Network (STDN) and the Tracking and Data Relay Satellite System (TDRSS). Both use radio frequencies within the so-called S-band from 2 to 4 gigahertz (one gigahertz = one billion hertz). The Tracking and Data Relay Satellite also uses the higher frequency Ku-band, from 13 to 16 gigahertz. The STDN system operates at low data rates; on the down-link from Orbiter to Earth, 192 000 binary digits (either a one or a zero), or 'bits', are transmitted per second, and on the up-link from Earth to Orbiter the rate is 72 000 bits per second. Much higher rates can be handled by the TDRSS, up to 50 million bits per second on the down-link, as well as analogue and television data. The ground station for the TDRSS is at White Sands, New Mexico, where the transmitted data may be recorded or sent to the NASA Johnson Space Center and the NASA Goddard Space Flight Center.

Two satellites of the Tracking and Data Relay Satellite System (TDRSS) will give communications coverage over most of a Spacelab flight since these geostationary satellites can usually be 'seen' from Spacelab. Data are sent to a TDRS which transfers them to Earth.

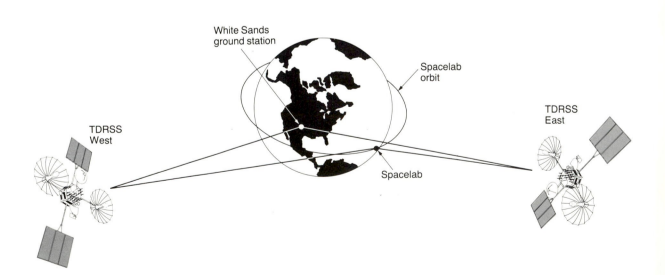

White Sands ground station

Spacelab orbit

TDRSS West

TDRSS East

Spacelab

The receiving antennae of the TDRSS ground station are at White Sands, New Mexico. The TDRSS will eventually replace NASA's worldwide network of ground stations.

Two geostationary satellites of the TDRSS should be stationed at longitudes 41 and 171 degrees West over the Atlantic and Pacific Oceans. About 85% of a normal Spacelab mission can thus be monitored. Each satellite has a mass of 2100 kilograms, and should operate for 10 years, with two large, 4.9-meter-diameter steerable antennae and three smaller ones. A solar array of span 17 metres provides 1.8 kilowatts of power. The satellites are stabilised in three axes and are launched by the Shuttle.

When Spacelab is 'out of sight' of the two satellites, data may be stored on-board Spacelab by the Spacelab High Data Rate Recorder (total capacity 30 billion bits at various rates up to 32 million bits per second) and the Orbiter Payload Recorder (capacity 3.4 billion bits at about 1000 bits per second). The necessary changes of magnetic tape reels are made by the crew.

The TDRSS and the STDN are linked to the Payload Operations Control Center (POCC) at Houston, where teams of ground-based experimenters can control their experiments and see real-time data being displayed. The experimenters can talk with the scientists on board, and see their experiments being conducted using a video link. The signals to and from the TDRSS and STDN are routed through the communications system of the Orbiter. The flow of data on board is controlled by the Command and Data Management Subsystem (CDMS), which is also concerned with data display, data recording, and the monitoring and control of the subsystems and experiments.

The CDMS has two parts. Low-speed data, at rates of up to 60 000 bits per second, are processed by the Data Processing Assembly (DPA), and high-speed data by the High Rate Data Assembly (HRDA). Using the Data Processing Assembly, commands are given to each experiment or subsystem via a Remote Acquisition Unit (RAU). Each Remote Acquisition Unit not only acquires and delivers data but also provides timing information and appropriate on/off commands. The signals are passed along a path called a 'data bus'. There are two such paths, an experiment data bus and a subsystem data bus for 'housekeeping' information. Also, experimental data and subsystem data are periodically transferred from the Remote Acquisition Units via an input/output unit which supervises the communications between the experiment and subsystem computers and the rest of the CDMS. The data are then processed by the computers for transmission, display or storage, either automatically or semi-automatically as directed by a crew member operating a keyboard.

This artist's impression of a TDRS shows two large, steerable antennae, a thirty-element S-band array, and smaller antennae which are used for radio communications in three different frequency bands (S, C and K).

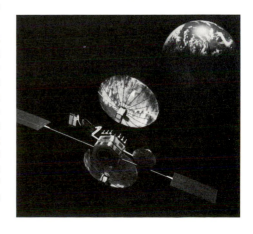

The TDRS-A and its Inertial Upper Stage rocket in the transport container on the way to the launch pad. There, the combination will be installed into Challenger's cargo bay for the STS-6 flight.

The keyboards used by the crew resemble normal typewriter keyboards. The alpha-numeric keys and the twenty-five function keys generate instructions that are recognised by the operating system's software. The associated Data Display Unit displays data, in green, yellow and red, which can be monitored and checked.

The high-rate data stream can be multiplexed to handle data from sixteen sources with individual data rates of up to 16 million bits per second, or up to 50 million bits per second from a single source. The data may be transmitted in real time or stored in the High Data Rate Recorder. Real-time transmission can be at rates of up to 50 million bits per second and recording can be at rates of up to 32 million bits per second. On the ground, the high-speed data go through processes, in a de-multiplexer, which are the reverse of those on board Spacelab in order to recover the information from the sixteen different sources.

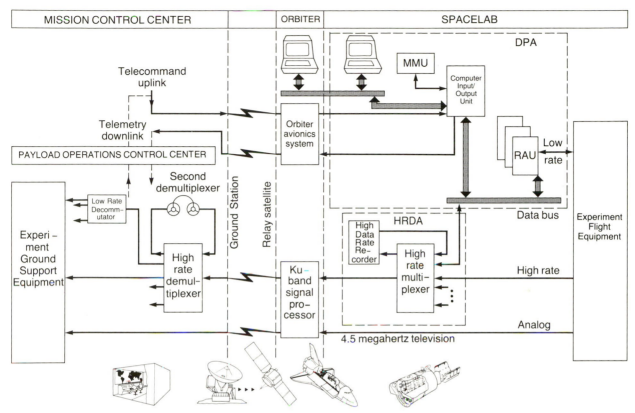

| MISSION CONTROL CENTER | ORBITER | SPACELAB |
| --- | --- | --- |

(Above) Data flow to and from the experiments is controlled by the Command and Data Management Subsystem. This diagram illustrates how the low-speed and high-speed data streams are handled via Spacelab, the Orbiter, the TDRSS and the Mission Control Center.

(Below) The scientist – astronauts use the keyboard and the Data Display Unit to conduct the experiments and to monitor their progress. Up to nine hundred and ninety-nine symbols can be displayed, on twenty-two lines. Two such units are normally contained in the Spacelab module.

Spacelab carries three identical computers; one is dedicated to the subsystems, one to the experiments, and the third is a 'spare'. Each computer is a general-purpose model, CIMSA 125 MS, with a 64k memory using 16-bit words.

Each computer is governed by software programs; each program consists of a set of coded instructions loaded into and stored by the 8 million-word Mass Memory Unit (MMU) before flight. This unit is a digital tape recorder which periodically loads the computer memories during the flight. All software programs, together with any necessary data tables and the desired sequence of events, are prepared and checked out as part of the experiment development and integration programme. Familiarisation with the data is an important aspect of each crew member's training.

For the payload, the overall operating system that controls such standard functions as data collection from the experiment Remote Acquisition Units is the Experiment Computer Operating System (ECOS); this accommodates subroutines related to each experiment (Experiment Computer Application Software, ECAS). ECAS must be specifically designed for each individual experiment, whereas ECOS is essentially the same for each flight, providing standard information from the Orbiter such as the time and navigational data. The most important feature of ECAS is its automatic control of the instruments; it will even provide the crew with a diagnosis if anything goes wrong. ECAS can also be used to manipulate target and observation files (for example, the coordinates of a star to be observed), and to perform any calculations needed to point an instrument in a particular direction.

Each Remote Acquisition Unit (RAU) is the clearing house for data associated with either experiments or subsystems. On/off commands, precise timing information and both experiment-generated and house-keeping data are directed as desired. Each RAU measures 17 × 42 × 18 centimetres and is mounted on a cold plate. The three RAUs shown on the right of this pallet are attached to their own cold plate, alongside two empty ones, and another supporting two more RAUs and an Experiment Power Distribution Box (lightly coloured).

The centralised data handling concept outlined here has taken a considerable time to prepare. In view of recent advances in micro-computers, computer tasks may be decentralised on future flights. In fact, the forthcoming German Spacelab mission, named D-1, will use specific processors to control groups of related experiments.

## The Spacelab Programme

European interest in space research immediately after World War II manifested itself in small national ventures or through collaboration with the Americans. In the early 1960s, scientists from a number of European countries discussed how they might combine their resources to use fully the many observing tools at their disposal to investigate the mysteries of space. As a result, the European Space Research Organisation (ESRO) was established; in 1975 the European Space Agency (ESA) was formed, combining the satellite work of ESRO with the rocket launcher activities which had previously been conducted by the European Launcher Development Organisation (ELDO). The present Member States of ESA are: Belgium, Denmark, France, the Federal Republic of Germany, Ireland, Italy, the Netherlands, Spain, Sweden, Switzerland and the United Kingdom. In addition, Austria and Norway are Associate Member States and Canada has special arrangements with the Agency.

Space research in Europe was initially concerned with the scientific investigation of space using unmanned satellites and then the exploitation of space using applications satellites, for example for meteorology and communications. European involvement in Man-in-space activities had to await the era of Spacelab.

In 1969, NASA invited ESA to participate in the development of an advanced Space Transportation System (STS); ESA accepted on behalf of its Member States. After studying possible options, and taking into account the political and technical aspirations of Europe, ESA chose as its contribution the manned orbital laboratory, to be carried in the cargo bay of the Orbiter.

The idea of Spacelab was formulated in 1971, but several studies had to be made before the final design took shape in 1973.

Hardware was developed in the middle and late 1970s, and Spacelab was delivered to NASA, for use with the Space Shuttle, in 1982.

Spacelab, being a cooperative venture, posed new and challenging management problems. The top-level intergovernmental agreement and key to this cooperative project is the Memorandum of Understanding (MOU) between NASA and ESA. Signed in Washington, D.C., in September 1973 by the Director General of ESRO (now ESA) and the Administrator of NASA, this document apportions the responsibilities between the Agencies. In essence, the Memorandum of Understanding states that ESA is responsible for:

(i) the design, development and manufacture of Spacelab and its associated Ground Support Equipment
(ii) the continuation of the capability for Europe to provide further Spacelabs
(iii) the availability of sustained engineering skills to meet the operational needs of NASA for the first few flights.

The extent of NASA's participation is:

(i) the provision of technical assistance as needed
(ii) the management of the operation of Spacelab as part of the STS
(iii) the development of certain items such as the access tunnel to Spacelab and a Spacelab simulator.

Spacelab was designed, developed and manufactured in Europe, with European funds. ESA ensured the overall management of the programme and represented Europe in discussions with NASA who provided valuable technical advice and support. The industrial team was led by MBB–ERNO at Bremen in the Federal Republic of Germany, and consisted of more than forty European companies in ESA Member States; the prime contractor is supported by ten co-contractors, themselves supported by a large number of European subcontractors.

The ESA/NASA Memorandum of Understanding on the Spacelab Programme is signed in Washington, D.C., by Dr A. Hocker (Director General, ESRO), seated at the far left, and Dr J.C. Fletcher (Administrator, NASA), on the far right, on behalf of their respective Agencies. In the centre are Minister C. Hanin (Chairman of the European Space Conference) and K. Rush (acting US Secretary of State).

The Spacelab Programme evolved during the 1970s. Hardware produced in the late 1970s flew for the first time in the early 1980s.

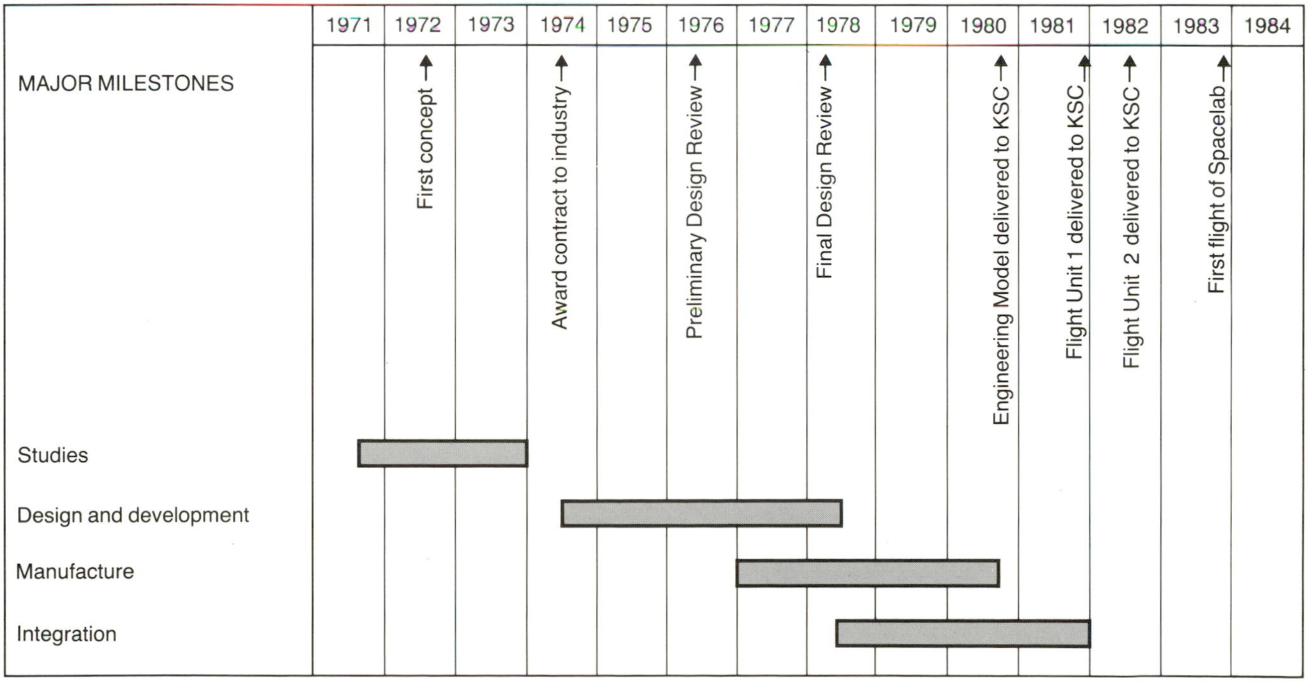
1971
1972
1973
1974
1975
1976
1977
1978
1979
1980
1981
1982
1983
1984
MAJOR MILESTONES
First concept
Award contract to industry
Preliminary Design Review
Final Design Review
Engineering Model delivered to KSC
Flight Unit 1 delivered to KSC
Flight Unit 2 delivered to KSC
First flight of Spacelab
Studies
Design and development
Manufacture
Integration

The total cost for the design, development, and manufacture of Spacelab is 760 million accounting units (MAU) or about £450 million (US$750 million). Funding has been arranged on an optional basis, with the largest contribution to ESA coming from the Federal Republic of Germany.

It is hoped that future use of Spacelab will be on an international basis, with experiments being supplied by the USA, Europe and non-contributing countries such as Japan and India. However, all the costs associated with experiments are excluded from the total cost specified for Spacelab development.

Early Spacelab studies (termed phase A and phase B) were conducted by three European consortia, COSMOS, MESH and STAR. These consortia consist of teams of industrial companies representing all of ESA's Member States. Considering the large contribution of the Federal Republic of Germany to the budget it was logical that the team to design and build Spacelab should be led by a German firm. Reflecting this philosophy, two teams, one led by ERNO and the other by MBB, the two principals of the MESH and COSMOS consortia, respectively, each defined a Spacelab concept independently. They studied in detail the design and cost of Spacelab. Based on their results, ESA's Council approved the Director General's recommendation in August 1973 that a viable Spacelab Programme could proceed within the cost constraints prevailing at that time. The development contract (phases C/D) was awarded to the team led by ERNO, but a later amalgamation of MBB and ERNO brought the Spacelab work of the two companies together again.

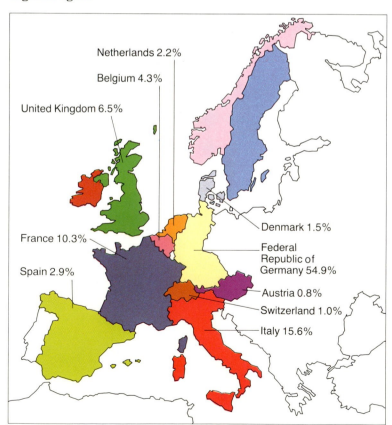

ESA's Spacelab Programme was an 'optional' one, in the sense that each Member State contributed according to its means and its interest in the manned space laboratory. Nine Member States and one Associate Member State participated, providing percentages of the total cost as shown.

(Left) The many parts that make up Spacelab were brought together in a special Integration Hall at ERNO's Bremen factory. The Engineering Model, shown on the left, was used during the development programme and will continue to be used to familiarise all those involved in the Spacelab Programme with Spacelab. The actual Flight Unit is pictured on the right. In the foreground, the igloo and several racks are shown in front of white-panelled racks containing a variety of electronic equipment.

## Members of the Spacelab team and their responsibilities

### *Prime contractor:*
MBB–ERNO (formerly VFW–Fokker/ERNO)
*Project management* ● *System engineering* ● *Product assurance* ● *Integration and check-out of mock-ups* ● *Engineering model and flight model*

### *Co-contractors:*
Aeritalia (Italy) *Module*

British Aerospace Dynamics Group (United Kingdom) *Pallet*

Fokker (Netherlands) *Airlocks and common payload support equipment*

Matra (France) *Command and data management*

AEG–Telefunken (Federal Republic of Germany) *Electrical power distribution*

Dornier Systems (Federal Republic of Germany) *Environment control and life support*

SABCA (Belgium) *'Igloo' and utility bridge*

Bell Telephone (Belgium) *Electrical Ground Support Equipment*

Kampsax (Denmark) *Computer software*

Sener (Spain) *Mechanical Ground Support Equipment*

(Below) Spacelab parts arrive at Kennedy Space Center, Florida, aboard a Lockheed C5A 'Galaxy'.

Vice-President George Bush shakes hands with ESA's Director General, Erik Quistgaard at the Spacelab arrival ceremony, Kennedy Space Center. Looking on are James Beggs (NASA Administrator, left), with Johannes Ortner (Chairman of ESA's Spacelab Programme Board) and Richard Smith (Director, Kennedy Space Center) at the podium.

To oversee the management of the Spacelab Programme, particularly the funding, a Spacelab Programme Board was set up by ESA. This Board, composed of representatives of the participating Member States, provided ESA's Director General and the Director of Space Transportation Systems (who directed the Spacelab Programme) with the necessary guidance, particularly concerning costs, for pursuing the Programme. Meetings of the board were held about every three or four months. Day-to-day management of the industrial effort was carried out by a Project Manager and a team of about a hundred scientists and technicians based at ESA's technology centre, ESTEC, at Noordwijk in the Netherlands.

To ensure a satisfactory liaison with NASA in carrying out the Memorandum of Understanding, a Joint Spacelab Working Group (JSLWG) was formed. The group contained five members from ESA and five from NASA, and was jointly chaired by the Directors of the ESA and NASA Spacelab efforts. Its principal task was to coordinate top-level actions which affected the interests of both Agencies. Scientists and technicians from both sides attended bimonthly meetings when issues relevant to their work were discussed.

Delivery of the Spacelab Flight Unit was planned for 1979 with the first launch set for 1980. Subsequent delays on both sides meant that the Spacelab hardware was eventually delivered in 1982. A Spacelab Flight Unit consists of a long module plus five pallets. In fact, the delivery of the Flight Unit (FU) occurred in two batches – FU1,

Spacelab in the Operations and Check-out Building at NASA's Kennedy Space Center, Florida. In the foreground is the Flight Unit module, with the Engineering Model behind, and two pallets on the right.

consisting of the long module and one pallet, in December 1981, and FU2, the remaining pallets and igloo, in July 1982. Each unit was accompanied by its associated software, spares and Ground Support Equipment such as jigs and electrical test gear. Previously, in December 1980, an Engineering Model – identical to the Flight Unit in design but not subjected to all the stringent testing required for space-flight – had been delivered. The Engineering Model was used by NASA and ESA to ensure that the correct level of proficiency for dealing with the Flight Unit had been achieved.

The design and development phase, in Europe, proceeded through the 1970s with each co-contractor and subcontractor producing its contribution to Spacelab. The final assembly of all these parts into one unit, and the test and check-out of this unit, was performed by the prime contractor, MBB–ERNO, at its Integration Hall in Bremen. This hall, with an area of 1125 square metres, was specially built to integrate and check-out the various items of Spacelab hardware and software. As part of the development process, several models of Spacelab were built as aids to the production of the Flight Unit itself. Thus, mock-ups were made to prove the lay-out and physical interfaces of the subsystem elements; an electrical test model was constructed to verify electrical connections and software. A further model was put together to check subsystems and the routing of cables and fluid lines. Finally, the Engineering Model was constructed; this was most valuable for verifying the manufacturing techniques used for the Flight Unit and for training. The Engineering Model, a very important part of the Spacelab Programme, is a fully functional model meeting nearly all the demanding requirements of the Flight Unit. Its integration in Europe helped to pave the way for the preparation of the Flight Unit itself. In the USA, it will be a significant aid in mission-planning, accommodation of equipment studies, and familiarisation with Spacelab.

The performance specification and correct interfacing of the various constituent parts of Spacelab are essential. Also Spacelab must interface correctly with the Space Shuttle. All these factors were assured by using a documentation system. This provided a master plan for the development programme, and supporting plans in a number of essential areas such as project control, design, safety, testing, software, manufacturing and procurement. In particular, documentation had to be issued to specify the exact requirements and interfaces that would exist for each hardware item built. The basic requirements start at the Spacelab system level; they grow and become more detailed, as appropriate, for each element making up the system. Although such documentation is often maligned as being costly and psychologically undesirable, there is no doubt that a logical and effective documentation system is essential for the proper control of a programme as complex as that of Spacelab.

To ensure that the control process was working efficiently and to identify potential problems as early as possible, the master plan scheduled a series of reviews. The most important of these were the Systems Requirements Review to ensure that the needs of experiments and the main Spacelab–Orbiter interfaces were met, the Preliminary Design Review to approve the basic design approach, and the Critical Design Review to authorise manufacturing to proceed. The reviews were attended by representatives from both ESA and NASA, plus engineers from the industrial companies concerned.

The tunnel for joining Spacelab to the Orbiter, manufactured for NASA by McDonnell Douglas.

The official emblem of the joint ESA/NASA Spacelab Programme symbolises international collaboration and Man performing research in Earth orbit.

The Spacelab Programme involved dedication, much hard work and some frustrations. Many technical problems arose which had to be solved, and schedules had to be revised for unforeseen reasons. The final product, however, is Spacelab, and the quality of design and workmanship embodied in this product of European engineering has been duly appreciated by NASA. On 13 January 1983, in Washington, D.C., Spacelab was certified as meeting all performance requirements and was, therefore, ready to take its place as a member of the Space Transportation System family.

The first Spacelab is a gift from Europe to NASA on the understanding that any future Spacelabs or parts of Spacelab needed by the USA would be purchased in Europe. Accordingly, NASA ordered from ESA a second Spacelab unit and associated spares. Some US$200 million will be made available to the industrial members of the Spacelab team.

The vital role played by NASA in this highly successful cooperative venture must be acknowledged. The Marshall Space Flight Center (MSFC) at Huntsville, Alabama, has been the leading NASA Center for monitoring the progress of the Spacelab Programme and for ensuring that Spacelab–Shuttle interfaces are compatible. Also, NASA has provided technical advice from its vast reservoir of space know-how. The other major NASA Centers concerned have been the NASA Headquarters, Washington, D.C., for policy and programme matters, the Kennedy Space Center in Florida for launch preparation, and the Johnson Space Center (JSC) in Texas for matters related to the flight itself. In addition, NASA has provided hardware for the Spacelab Programme, including the 6-metre-long tunnel used for access to the Spacelab module from the Orbiter, de-multiplexers for interpretation of Spacelab-transmitted data, instrumentation to monitor the performance of Spacelab on its early flights, crew training facilities at MSFC and JSC, and the Experiment Computer Operating System software. Together with their European counterparts, the NASA representatives have ensured that an international venture in advanced technology can be harmoniously carried out and successfully concluded.

The completion of the Spacelab Programme has clearly given ESA and the aerospace industries of its Member States much very valuable experience in the management of a large-scale international enterprise. The technologies which have been built up in Europe can contribute to further space programmes, both manned and unmanned. The ground work has surely been laid for further large projects on which different nations can work well together.

## The Space Transportation System

The Space Shuttle Program has tremendous potential for mankind. It will transform the frontiers of space into easily accessible and familiar territory. The Space Shuttle can put objects into orbit around the Earth at comparatively low altitudes, between 200 and 600 kilometres. In space jargon, the Shuttle is used to inject a payload into a low Earth orbit. Shuttling repeatedly from Earth into space, and back again, the Shuttle is the basis of the advanced Space Transportation System (STS). It is NASA's workhorse of the future. The STS family includes not only the Shuttle but also Spacelab and additional rockets, termed upper stages, for putting payloads into higher orbits still or orbits of different inclinations. The Tracking and Data Relay Satellite System, and the ground stations needed for the control and tracking of STS members in orbit, complete the family.

# INTO ORBIT ABOARD THE SPACE SHUTTLE

*'With the first orbital flight of the Space Shuttle, the curtain rises on an era that will shape US space exploration for the next decade and perhaps for the remainder of the century.'*

Adlai Stevenson

The Space Shuttle blasts off for the first time in April 1981 from Kennedy Space Center, Florida.

For higher orbits, a Payload Assist Module (PAM) or an Inertial Upper Stage (IUS) rocket is attached to the satellite and carried up in the cargo bay of the Shuttle Orbiter before being released in low Earth orbit. At a safe distance from the Shuttle, the rocket engine is fired and the satellite is carried into the desired orbit at the required inclination. One such high-altitude orbit is the commercially important geostationary orbit, at an altitude of about 35 800 kilometres. There, the orbital period of the satellite matches that of the Earth, and the satellite appears to remain over one particular place on the Earth's surface. Such geostationary satellites are essential to the Tracking and Data Relay Satellite System used to transmit radio signals from the Orbiter or its payload back to Earth.

Space vehicles may be launched and operated in a most cost-effective way using the STS family. Shuttle payloads may consist of satellites for scientific or technological purposes, or any equipment designed to operate in orbit (that is for so-called on-orbit operation). Spacelab, with its scientist–astronauts in low Earth orbit, is a very special payload of the Space Shuttle. Working in conjunction with the Shuttle Orbiter, it provides what might be described as a laboratory in space, or a short-duration Space Station. It fits within the large cargo bay of the Orbiter, and remains there throughout the flight in space.

The STS will be used throughout the 1980s and 1990s and, possibly, into the early part of the twenty-first century. Spacelab, as an integral part of this system, has therefore an important role to play in the future of manned spaceflight.

## Re-usable space equipment

Satellites are usually launched by expendable rockets, with the rocket engines and protective shields being jettisoned into space. However, the major elements of the Space Shuttle have been designed to be recovered and used again; the Shuttle is a re-usable system. The Orbiter itself may be used up to a hundred times, being launched vertically like a rocket and landing like a glider. The solid-fuel booster rockets are recovered from the sea, refurbished and refuelled. The only part that cannot be re-used is the large external tank which carries the ascent fuel.

Spacelab is itself re-usable; up to fifty reflights are possible. On its return to Earth it may be fitted out with a new set of instruments, or the existing equipment may be modified for its next flight into space. In this way, a completely new scientific topic may be investigated, or a further assessment may be made of the same phenomenon but with somewhat different techniques. Furthermore, the recoverability of Spacelab means that data records – such as photographic film, magnetic tapes, samples or specimens – are easily accessible to research scientists.

One particular advantage of re-usability is readily apparent. Since expensive equipment is not thrown away, the cost of launching and maintaining satellites can be kept lower than those associated with expendable systems. Re-usability and versatility are complementary advantages of the Space Shuttle over conventional launching methods.

# The Space Shuttle

The Space Shuttle is extremely versatile; it can handle large weights and volumes. Because it is re-usable it does this at relatively low cost.

At launch, the Space Shuttle consists of the Orbiter vehicle, together with a large external tank which contains liquid oxygen and liquid hydrogen for fuelling the Orbiter engines during ascent, and two solid-fuel rocket boosters alongside the tank. These boosters provide additional thrust during the first 2 minutes of flight. The ensemble has a mass of about 2 million kilograms (2000 tonnes) at launch and measures about 56 metres overall, about half the length of the Saturn-V rocket used for the Apollo Program. It has the capability for placing almost 30 000 kilograms into a circular orbit at about 200 to 400 kilometres altitude, for the case of a due-East launch from the Kennedy Space Center at Cape Canaveral, Florida. This is equivalent to putting a loaded 'juggernaut' lorry into low Earth orbit. From Vandenberg Air Force Base, California, the Space Shuttle will put payloads of up to almost 18 000 kilograms into a polar orbit (near 90 degrees inclination) at similar altitudes.

The heart of the Space Shuttle is the 85 000-kilogram Orbiter built by Rockwell International. It is about the size and weight of a McDonnell Douglas DC-9 aeroplane, and is home for up to eight crew members whilst in orbit around the Earth. Although the Orbiter takes off like a rocket, it is shaped like an aeroplane (37 metres long, with a wing span of 24 metres). The Orbiter glides to its landing point using its aerodynamic controls and its small reaction and control rocket engines. The payload bay is about 18 metres long and 4.6 metres in diameter, large enough to house a Greyhound bus.

The Orbiter itself has three main Rocketdyne engines with a total thrust of 6.3 million newtons. Of original design, with special pumps and turbine blades, these rocket engines burn liquid oxygen and liquid hydrogen. Some 2 million litres (520 000 US gallons) of propellants are contained in the large external tank, of length 47 metres, and diameter 8.7 metres. This tank acts as the structural backbone of the Shuttle during launch.

It takes only 8 minutes for the 700 000 kilograms of propellant to burn. The liquid oxygen and liquid hydrogen propellants must be maintained at about −147 and −251 degrees Celsius, respectively, to prevent boiling and hence the loss of fuel. Propellants which must be maintained at low temperatures are termed cryogenic; tanks containing such super-cold fluids must be specially designed. Here, a spray-on insulation is used to make sure that the propellants remain cool during flight and to reduce the formation of ice on the tank surface. Some of this ice is vibrated free during launch and may hit the cockpit windows.

The boost thrust obtained from the two solid-fuel rockets, which together burn 1 million kilograms of aluminium perchlorate and aluminium, amounts to 24 million newtons at sea level. Each Thiokol rocket booster measures 45 metres in length and has a diameter of 3.7 metres. The solid-fuel rocket boosters are fixed to the large tank by special thrust attachments. The three main engines of the Orbiter and the two booster rockets together produce almost as much thrust as did the huge Saturn-V rocket of the Apollo Program.

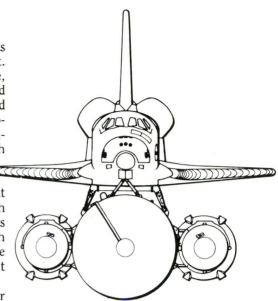

(Above) The relative size of the external fuel tank is evident in this head-on view of the Space Shuttle.

(Below) For launch, the Orbiter rides piggy-back on the large external fuel tank. The two solid-fuel rockets fixed on either side of the tank produce 80% of the thrust at lift-off.

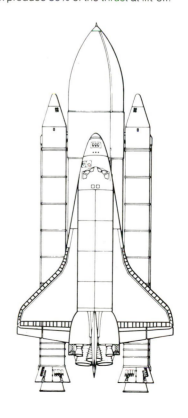

The Orbiters are built by Rockwell International in California, using many standard methods of aircraft construction.

A thrust of 54 000 newtons is produced by two smaller rocket engines near the tail of the Orbiter. Monomethylhydrazine (the fuel) and nitrogen tetroxide (the oxidiser), which are stored on-board, ignite on contact; such materials are said to be hypergolic. These Aerojet rocket engines form the Orbital Maneuvering System (OMS) which finally puts the Orbiter into orbit and which makes small corrections to this orbit. Manoeuvres needed to ensure that the Orbiter is in the desired orbit at the correct height (both apogee and perigee) and to make minor changes in inclination are carried out using the OMS at predetermined times. The engines burn for a minute or two at a time to change the Orbiter's speed by about 50 metres per second.

The Orbiter is orientated in space using any of thirty-eight primary and six secondary (for fine, or vernier, control) thrusters mounted forward and aft. The propellants used here are also monomethylhydrazine and nitrogen tetroxide, but only small thrusts are generated (3900 newtons for primaries, 110 newtons for verniers). This Reaction Control System (RCS) is used to change the orientation of the Orbiter, and to maintain the attitude (that is, the position of the Orbiter with respect to the vertical) fixed as the astronauts move about the Orbiter. When the Orbiter is in the nose-down (or tail-down) attitude the gradient of the Earth's gravitational field tends to stabilise it in that position.

The Orbiter is made mainly of aluminium alloy which becomes weakened at temperatures above 175 degrees Celsius. During re-entry into the Earth's atmosphere, temperatures are generated which considerably exceed even the melting point of aluminium (660 degrees Celsius), so the vehicle needs to be thermally insulated. Low-density, rather brittle, white ceramic (silica) tiles are bonded to the surface of the Orbiter. With their low thermal conductivity, they act as an insulator between the hot air and the aluminium. These tiles radiate heat into space, away from the Shuttle structure and may reach temperatures of up to 1200 degrees Celsius on their outer surfaces. The tiles do not evaporate, or ablate, as did the heat shields of earlier spacecraft which returned to Earth.

A technician fits a heat-resistant tile to the underside of the Orbiter. Each tile is individually shaped to match the surface contours, and its thickness depends on its position.

The Space Shuttle Columbia is prepared for launch from pad 39A at Kennedy Space Center.

Some 30 000 individually machined tiles are used; these vary in thickness (from 0.25 to 15 centimetres) depending on their locations. Besides high temperatures, these tiles must withstand aerodynamic pressure gradients, shocks and buffeting. The development of this thermal protection system posed considerable problems, which are well known due to the much publicised missing tiles of early Shuttle flights. The current system is perfectly effective, but improved, lower-weight materials will be used in later flights. The tiles cover two-thirds of the Orbiter's surface, the remaining, mainly upper surface, area being covered by sheets of insulating felt. These can resist temperatures of up to 400 degrees Celsius.

## Mission profile

The Space Shuttle is launched vertically. The liquid-fuel engines of the Orbiter are ignited 6.6 seconds before lift-off. When the solid-fuel engines are ignited, the Space Shuttle rises majestically from the launch pad. At lift-off, the solid-fuel rocket boosters and Orbiter main engines burn in parallel developing a total thrust of some 30 million newtons. They propel the Space Shuttle, with a smooth acceleration of up to only 3$g$, through the sound barrier after only 16 seconds. The Shuttle is then travelling at 'Mach 1', a speed of 0.34 kilometres per second, a little over 1200 kilometres per hour. When the solid fuel is exhausted the rockets stop firing, about 125 seconds into the flight; this is called burn-out. The Orbiter is then moving at four times the speed of sound, 'Mach 4', at a height of 45 kilometres. The two solid-fuel rockets are jettisoned over the sea to descend by parachute. A special retrieval ship recovers them about 300 kilometres down range and tows them back to Kennedy Space Center for refurbishment.

The Orbiter continues its ascent on its three main engines using fuel from the large external tank. The engines operate for a further 6 minutes and are then shut down. About 20 seconds later, the large tank, which is made of aluminium alloy 5 centimetres thick, separates at a height of 120 kilometres, re-enters the atmosphere and burns up. Its remnants fall into the Indian Ocean about 20 000 kilometres down range. The separation takes place just before the vehicle reaches orbital speed; the Shuttle is injected into the desired orbit 2 or 3 minutes later using the Orbital Maneuvering System engines.

Once the Orbiter is safely in orbit, the on-orbit operations commence with the opening of the doors of the cargo bay. The doors resemble a huge clamshell, and are designed to open and close in the zero-g environment; this operation, which is effortless in space, is quite a major undertaking on the ground. Inside the doors are radiators which must be exposed to space to dissipate the Orbiter's excess heat.

When the cargo of the Space Shuttle consists of Spacelab, the required orientation is provided by the Orbiter. Spacelab is checked out and opened up, and the scientist–astronauts enter by 'swimming' along the tunnel to commence their duties. The mission may last a week or more, with the crew working in Spacelab in shifts and spending their off-duty hours in the Orbiter. Flights may involve Extra-Vehicular Activity (EVA), where the crew work outside the Orbiter in spacesuits.

The main events of a Space Shuttle mission are illustrated here. After launch, various activities can be carried out in space. Then the Shuttle re-enters the Earth's atmosphere and lands at its home base. The next payload to be flown joins the Orbiter in the Orbiter Processing Facility. After the complete Space Shuttle has been put together in the Vehicle Assembly Building, the Orbiter is ready for another launch.

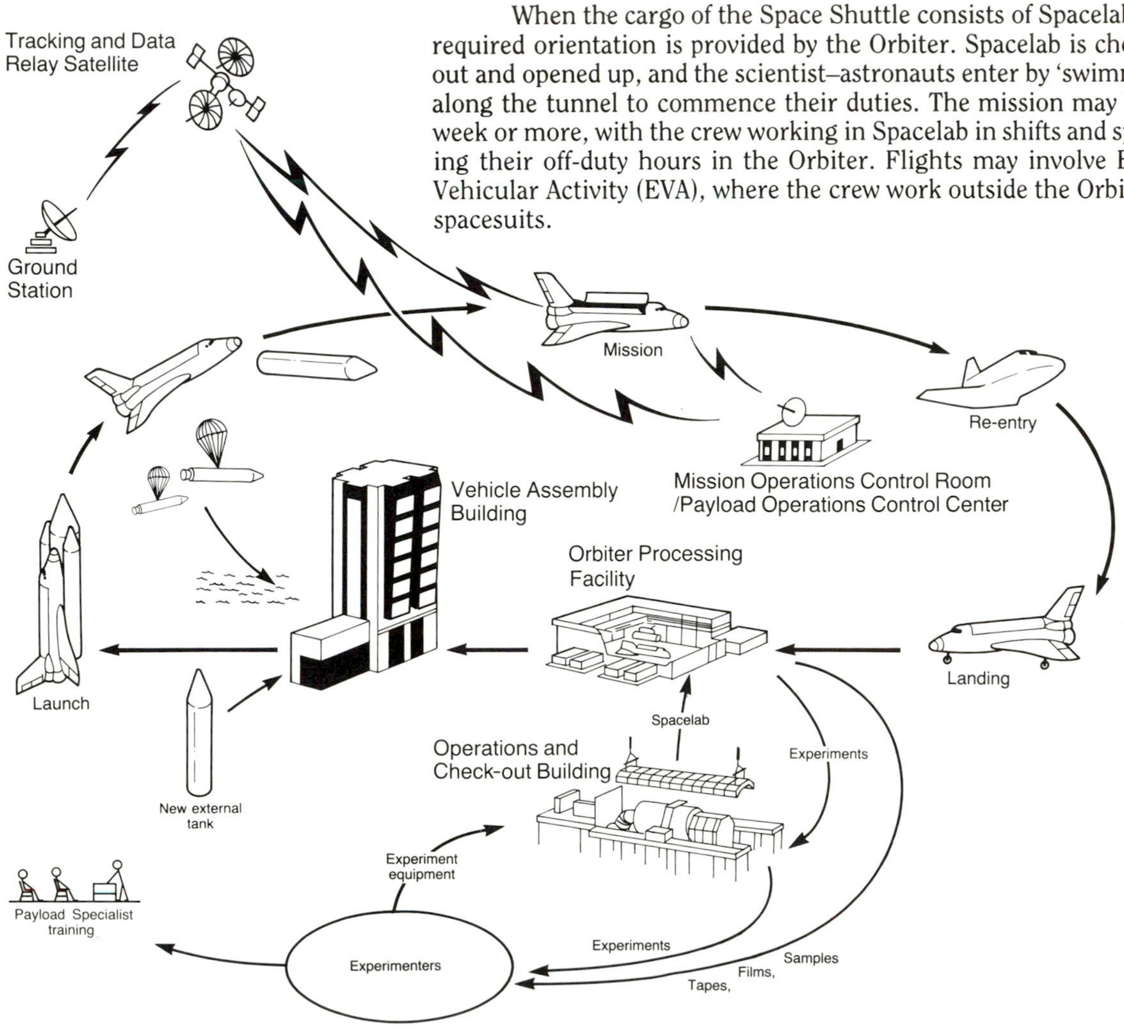

Tracking and Data Relay Satellite

Ground Station

Mission

Re-entry

Mission Operations Control Room /Payload Operations Control Center

Vehicle Assembly Building

Orbiter Processing Facility

Landing

Launch

New external tank

Spacelab

Experiments

Operations and Check-out Building

Experiment equipment

Payload Specialist training

Experimenters

Experiments

Samples

Films, Tapes,

In order to take payloads out of the cargo bay or to move an object on the pallet, a Remote Manipulator System (RMS) is used. A remotely controlled manipulator arm, with mass 400 kilograms and length 15 metres, was developed in Canada. The position of the arm, which has 'elbow' and 'wrist' joints, is controlled by astronauts from the Orbiter's aft flight deck using information from television cameras mounted on the arm and in the cargo bay. The arm may also be used for retrieving satellites for return to Earth. The maximum payload that can be safely brought back is 14 500 kilograms. However, neither the RMS nor EVA was used on the first Spacelab flight.

Three electrolytic fuel cells, in which hydrogen and oxygen are combined to make water, produce 21 kilowatts of electricity, besides supplementing the crew's supply of drinking water. On later flights of Space Shuttle, extra power could be provided by a further fuel cell or by an array of solar cells. Other on-board subsystems control the environment within the Orbiter, and provide for data management. A set of five identical, and interacting, IBM computers performs all the tasks needed for flight control and mission execution.

When the orbital part of the mission is completed, the final task is to close the cargo bay doors. The Orbiter is then turned round so that the Orbital Maneuvering System engines face the direction of flight. The engines are retro-fired for 2 or 3 minutes, which reduces the speed of the Orbiter (7.9 kilometres per second) by 0.1 kilometre per second. The Orbiter is turned round again, and re-enters the atmosphere nose first. The initial re-entry is made at a high angle of inclination (up to 40 degrees) to the air stream. During its flight through the atmosphere this angle of attack is reduced, and use is made of the aerodynamic surfaces. High temperatures are generated due to air friction during this 9-minute phase. Temperatures may reach 1600 degrees Celsius where the air is brought completely to rest at the nose and leading edges (known as stagnation points), and 300 degrees Celsius on the leeward surfaces. However, the silica tiles over most of the Orbiter's body and reinforced carbon-fibre inserts at the nose and leading edges keep the temperature of the Orbiter's aluminium inner structure well below its operating value of 175 degrees Celsius. The high temperatures around the Orbiter during re-entry ionise the atmosphere. High electron densities in the vicinity of the Orbiter and its radio antennae, much higher than in the ambient ionosphere, prevent radio communication even at the highest frequencies, causing a period of radio blackout between the Orbiter and Mission Control.

During its re-entry flight of about 8000 kilometres the Orbiter reduces speed from 7.8 kilometres per second to almost zero (actually 0.1 kilometres per second). Thus the vehicle experiences a variety of aerodynamic regimes from hypersonic, through supersonic, to subsonic. The rate of descent is about ten times greater than that experienced when landing in a jet airliner. The Orbiter lands like a conventional glider, at a speed of about 0.1 kilometres per second (185 knots). The computer-aided final approach must be exactly on target since the Orbiter has no engines to make a circuit of the landing strip and attempt another landing. However, by banking the Orbiter during descent, it may land up to 1600 kilometres either side of the nominal re-entry track. This banking also helps the Orbiter to lose energy. The total time for re-entry, from retro-firing the OMS engines to landing, is about 55 minutes.

The Orbiter's OMS engines are fired during the STS-7 flight.

After the STS-3 flight, Columbia lands at Northrop Strip, New Mexico.

An aerial view of Kennedy Space Center showing the large Vehicle Assembly Building, the twin Orbiter Processing Facilities and, in the distance, the Space Shuttle runway. On the right is the original Apollo launch tower which is being dismantled.

The runway at Kennedy Space Center, normally to be used for Orbiter landings, is a 4.5-kilometre-long concrete strip. The first Spacelab landing site, however, was at the Edwards Air Force Base in California. This site was used because of the lack of an advanced approach and landing aid known as a 'head-up' display in the Orbiter Columbia. This display will permit the pilot, who has 180-degree-wide view, to perceive important landing information in visual form without his moving his head and thereby being distracted from the job in hand. Future Orbiters will be fitted with this aid so that landing at Kennedy Space Center will become routine. This landing of the first Spacelab at a remote site entailed special post-landing operations to protect the payload, and the later return of the Orbiter and its precious cargo to Kennedy Space Center atop a specially adapted Boeing 747.

Once the Orbiter is again on terra firma, it is refurbished and repaired; payloads are unloaded and reloaded in the Orbiter Processing Facility (OPF), purpose-built at the Kennedy Space Center. Here the Orbiter remains in a horizontal position until it is ready for flight again, when it is towed to the nearby Vehicle Assembly Building (VAB) where the various elements of the Space Shuttle are assembled. The VAB, built for the Apollo Program, was similarly used for integrating the various stages of the Saturn-V and Command and Service Module of the Apollo system.

Inside the VAB, the Orbiter is swung into a vertical position, and the mated solid-fuel rockets and external tank are attached. The external tank is new from its Martin Marietta factory in Mississippi, but the solid-fuel rockets have been recovered from the sea after a previous flight and recharged with fuel. The Space Shuttle ensemble, as tall as Nelson's Column in Trafalgar Square, is then transported some 5 kilometres to launch pad 39A (which was originally used for the Apollo Moon shots) upon the world's largest tracked vehicle. Powered by two 2750 horsepower diesel engines, this crawler vehicle has a top speed of 1.6 kilometres per hour. Its fuel consumption is 500 litres per kilometre or 0.007 miles per gallon! After a 'turnaround' time that could eventually be as short as 2 weeks, all is now ready for another launch.

The launch itself is controlled remotely from the Launch Control Center situated next to the VAB. Check-out and countdown, and other operations associated with the launch, are performed automatically by means of computers of the Launch Processing System.

Challenger is taken to the launch pad for the seventh STS flight.

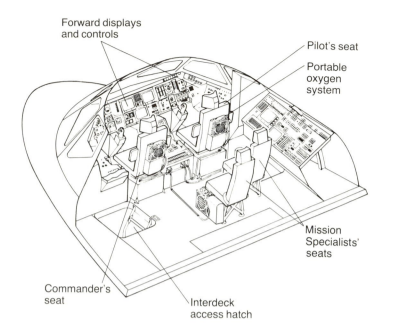

Forward displays
and controls

Pilot's seat

Portable
oxygen
system

Mission
Specialists'
seats

Commander's
seat

Interdeck
access hatch

Cut-away diagram of the flight deck from where
the Commander and Pilot control the Orbiter
during the flight.

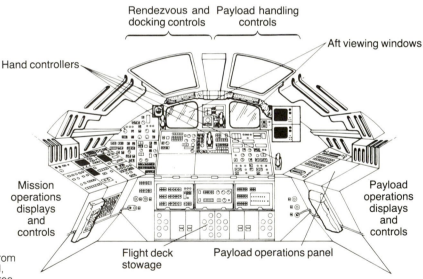

Rendezvous and   Payload handling
docking controls      controls

Aft viewing windows

Hand controllers

Mission
operations
displays
and
controls

Payload
operations
displays
and
controls

Flight deck
stowage

Payload operations panel

The crew can conduct payload operations from
the aft flight deck. When Spacelab is carried,
its subsystems can be controlled from the area
shown on the left, whereas experiments can be
operated from the right-hand side of the aft
flight deck.

## Living in the Orbiter

The crew and passengers of the Orbiter occupy a two-level cabin of
over 70 cubic metres within the crew module at the forward end of the
vehicle. As with Spacelab, the cabin is pressurised and filled with air
whose temperature and humidity are kept within comfortable limits.
The upper section, or flight deck, contains the controls and displays
needed for flying the Orbiter and controlling the payload. Four crew
can be seated here. The mid-deck level, reached through an open
hatch, is where the crew can live and relax. There are normally three

seats there for the scientist–astronauts to use during launch and re-entry.

On the left of the mid-deck area is the kitchen, or galley. This is equipped with a pantry, an electric oven, a water heater and trays on which food is served. Pure water made in the Orbiter's fuel cells, which produce electricity, is added to freeze-dried food – all food is kept in plastic packages or cans, sufficient for a single portion. Soft drinks, tea and coffee are made in the same way by adding water to powder contained in a plastic package.

All the food selected for a Shuttle flight is not only nutritious, tasty and digestable but also compact, light in weight and easy to prepare. Three meals, with planned – yet varied – menus, provide each astronaut with sufficient calories per day. Meals are enjoyed around a table in the mid-deck area. Rubbish and the empty packages are put into a plastic bag, which is then sealed for return to Earth. This bag is put through a hatch into the lower deck – besides providing storage space this area houses the Orbiter's Environment Control Subsystem.

Curtained off from the left-hand side of the mid-deck area are a washroom and toilet. Water is drawn into a small basin by a stream of air for washing. A crew member may shave using a normal razor and shaving cream or a clockwork-wound razor. The toilet, designed for use by both men and women either standing or sitting, also utilises a flow of air to aid removal of the body's waste products.

On the right side of the mid-deck area are three conventional bunks, or 'sleep stations', and also a vertical sleep station. This is perfectly comfortable in the weightless conditions. Each crew member has a locker for his personal belongings there too.

The three bunks can be removed and extra seats installed if a second Space Shuttle has to rescue crew members of an earlier mission. The second mission with a Commander, Pilot and Mission Specialist could rescue up to eight astronauts and scientists.

An astronaut, wearing simple life-support equipment, is cramped in a plastic sphere, one metre in diameter. Within this ball, he could be rescued by a spacesuited Mission Specialist.

The Orbiter Enterprise is taken aloft by a
Boeing 747 to determine its gliding and landing
characteristics.

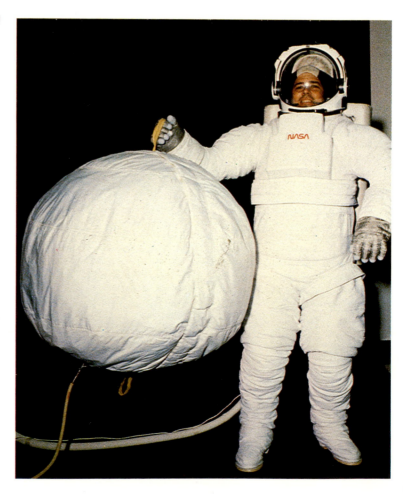

A spacesuited Mission Specialist holds up one
of the spheres in which an astronaut could be
rescued.

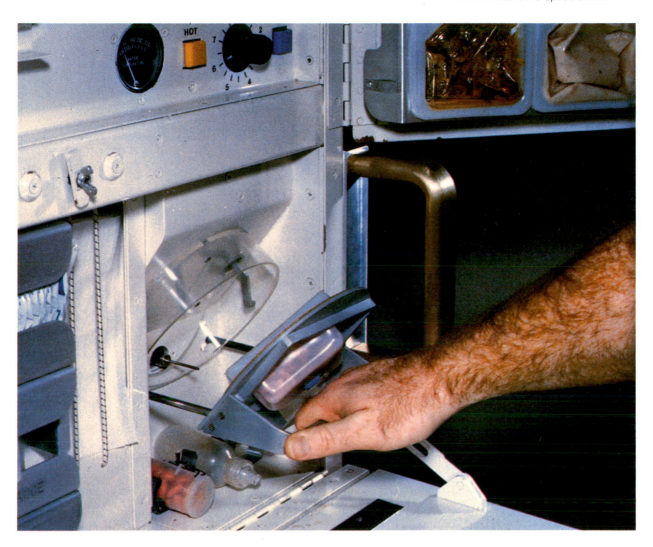

In the galley, food in a plastic container is about to be injected with water.

An astronaut's main meal is not only nourishing but easy to eat and digest.

On its first flight, Columbia's vast cargo bay contained only instruments to give information on the flight. A few tiles are seen to be missing from the pod of the Orbital Maneuvering System.

The OSTA-1 payload, mounted on an Engineering Model pallet, is lowered into the cargo bay of Columbia (STS-2). Very evident is the large antenna of the Shuttle Imaging Radar experiment.

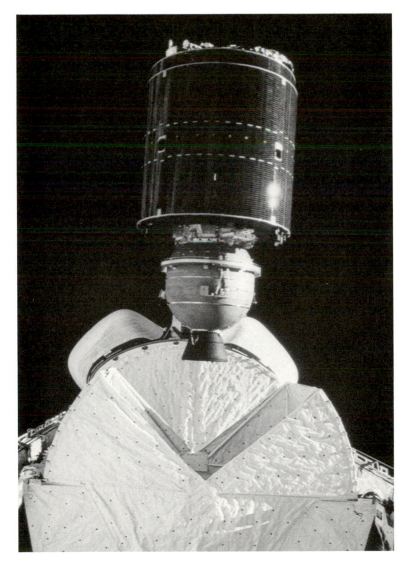

The Shuttle Pallet Satellite (SPAS) is soon to be released by the remote manipulator arm to fly freely in space near Challenger (STS-7).

The Satellite Business System's satellite (SBS-3) was the first commercial satellite to be launched by the Space Shuttle (STS-5).

First into space aboard the Shuttle were John Young and Robert Crippen.

# Space Shuttle activities to date

The re-usable Space Shuttle was developed under the auspices of NASA, with Rockwell International as the prime contractor, at an overall cost of US$10 billion. Although the first manned orbital flight was originally planned for 1979, the radical nature of the technology involved led to delays in the Program and the first launch was 2 years late. Particular problems were encountered with the Orbiter's main rocket engines, which operate at much higher pressures than those associated with the Saturn series, and the notorious tiles of the thermal protection system. The roll-out of the first Orbiter, designated number 101 and named Enterprise, took place in September 1976. Its flying characteristics at low speeds were established in the approach and landing tests after spring-ejecting it from a piggy-back position on a Boeing 747. These tests were successfully completed from Edwards Air Force Base in California by October 1977.

Meanwhile the development of other elements of the Shuttle Program continued. The structure of the large expendable tank was tested in early 1978, and the solid-fuel rocket boosters were ready in July 1979.

After structural tests carried out on Enterprise, this Orbiter was shipped to Kennedy Space Center where on-the-pad tests at the actual launch site were performed in April and May 1979. These tests ensured that all the geometric and functional interfaces between Space Shuttle and the Ground Support Equipment were compatible. The Orbiter main engines were fully checked during 1980. Assembly of the Space Shuttle 'stack' for its first flight started in January 1980, with the erection of the solid boosters in the VAB. The Orbiter Columbia arrived at Kennedy Space Center in March 1979 and, after the tiles were installed, it joined the Shuttle stack in November 1980. Static firing tests of rocket engines were conducted on the pad early in 1981.

After a delay of 2 days, due to a minor on-board problem of synchronisation between computers, the spectacular, yet comparatively gentle, launch of Columbia took place on 12 April 1981. Thus the first flight of the Space Shuttle coincided exactly with the twentieth anniversary of Man's first spaceflight by Yuri Gagarin. For this STS-1 mission, which lasted 54 hours, the crew were John Young (Commander) and Robert Crippen (Pilot). They became the first men to fly the Space Shuttle into orbit (at 40 degrees inclination) and, after completing 36 orbits, returned it safely to Earth on a dry lake bed at Edwards Air Force Base, California. Although the cargo bay was empty, instruments monitored the performance of the Shuttle systems in space and during the Orbiter's glide through the atmosphere. Seventeen of the protective tiles covering the engine housing became loose or unstuck during the vibration at launch. Nonetheless, the first Space Shuttle flight was described by John Young as being 'much better than anyone ever expected. This is the world's greatest flying machine.' The first-shift Flight Director exclaimed 'It's absolutely amazing. We have a super vehicle up there.' The era of re-usable space hardware had begun.

In all, a fleet of four space Orbiters will be used; its members are Columbia, Challenger, Discovery and Atlantis. Production of the last two cost a further US$5.6 billion. A fifth Orbiter will be built if fully justified and if funds become available. Enterprise is used only for ground and airborne tests and for demonstrations.

Some fifty launches have been scheduled from the Kennedy Space Center, until 1987, with an additional eight launches from Vandenberg Air Force Base, California. A variety of tasks will be performed during the corresponding missions, including four Spacelab flights, the launching of communications satellites, deep-space probes and Earth-orbiting satellites, US Department of Defense applications, and satellite servicing and repair in orbit. In mid-1982, NASA announced that the cost of a Space Shuttle launch would be US$71 million between 1985 and 1988. This is equivalent to US$2400 per kilogram of payload, for the maximum payload achieved with a launch due east from Kennedy Space Center.

## Orbiter flights subsequent to STS-1

| Flight number | Launch date | Orbiter | Crew | Duration of flight (days) | Highlights |
|---|---|---|---|---|---|
| STS-2 | 12 November 1981 | Columbia | Joe Engle<br>Dick Truly | 2 | Check-out system and perform experiments (OSTA-1) |
| STS-3 | 22 March 1982 | Columbia | Jack Lousma<br>Charles Fullerton | 8 | Check-out system and conduct extended mission. Perform scientific experiments (OSS-1) |
| STS-4 | 27 June 1982 | Columbia | Thomas Mattingly<br>Henry Hartsfield | 7 | Launch of US Department of Defense payload |
| STS-5 | 11 November 1982 | Columbia | Vance Brand<br>Robert Overmyer<br>Joe Allen<br>William Lenoir | 5 | Conduct first operational flight<br>Place Satellite Business System SBS-3 and Canada's Anik C-3 in orbit |
| STS-6 | 4 April 1983 | Challenger | Paul Weitz<br>Karol Bobko<br>Story Musgrave<br>Donald Peterson | 5 | Launch of TDRS-A/IUS.<br>Make first EVA |
| STS-7 | 18 June 1983 | Challenger | Bob Crippen<br>Fred Hauck<br>Sally Ride<br>John Fabian<br>Norman Thagard | 6 | Place Indonesian Palapa-B and Canada's Anik C-2 in orbit<br>Deploy and retrieve European SPAS<br>Fly first US female astronaut |
| STS-8 | 30 August 1983 | Challenger | Dick Truly<br>Dan Brandenstein<br>Guion Bluford<br>Dale Gardner<br>William Thornton | 6 | Place Indian Insat-1B in orbit<br>Test TDRS-A<br>Make first night time launch and landing |
| STS-9 | 28 November 1983 | Columbia | John Young<br>Brewster Shaw<br>Robert Parker<br>Owen Garriott<br>Byron Lichtenberg<br>Ulf Merbold | 10 | First Spacelab flight |

# Achievements of STS-2 and STS-3

Packages of experiments provided by the NASA Office of Space and Terrestrial Applications (OSTA) and the NASA Office of Space Sciences (OSS), numbered OSTA-1 and OSS-1, were taken up on the second and third Orbiter flights, respectively. Of particular interest were the Synthetic Aperture Radar (SAR) pictures of OSTA-1 and the sensing of the Orbiter environment and the performance of plasma investigations by means of the Plasma Diagnostics Package (PDP) on OSS-1. The majority of experiments on both packages were mounted on a Spacelab pallet. Hence the STS-2 mission represented the first time that hardware manufactured in Europe flew with the new STS.

Various factors delayed the flight of STS-2. Propellant spillage damaged almost 400 tiles, and a further 350 had to be replaced. There was also a fault in the power unit for the Orbiter's hydraulic system. Finally the launch was delayed by bad weather. The mission itself was shortened to 36 orbits (at 38 degrees inclination) because of a fuel cell problem. Operation of the remotely controlled manipulator arm with its shoulder, elbow and wrist joints was, however, successfully demonstrated. On return to Earth at the NASA Dryden Flight Research Center/Edwards Air Force Base, California, it was noticed that the upper surfaces of some tiles had been removed. It is believed that water from showers of rain had soaked into them while Columbia was waiting on the launch pad, and had turned to steam under the high temperatures of re-entry. But the tiles were soon repaired, and the re-usability of space hardware had, nonetheless, been effectively demonstrated.

Five groups of experiments were carried on the pallet for STS-2. The largest of these was a microwave radar (1.3 gigahertz, or 23 centimetres wavelength) based on the successful, though short-lived, Seasat satellite SAR design. This provided map-like images of the Earth's surface in relief, with 40 metres resolution, in which objects larger than 40 metres were distinguishable. This experiment was able to operate at night and under cloudy conditions, from an altitude of 262 kilometres. Eight hours of data were obtained over 10 million square kilometres of North, Central and South America, Southern Europe, Africa, Indonesia and Australia. An unexpected result was the discovery in the Eastern Sahara of river valleys in which water flowed thousands of years ago. Because the moisture content of the desert soil is so low, the radar waves penetrated to depths of a few metres before being reflected. These ancient features are covered by drift and sand dunes. Similarly, centres of human habitation during the Stone Age have been discovered near ancient river channels. Other results of the SAR were foreseen, namely the mapping of geological features such as faults, folds, fractures or rocky outcrops. Volcanic lava flows and also craters produced by meteorite impacts one million years ago appear bright on the radar image because of their surface roughness on a scale of about 10 cm. Brightness variations delineate cultivated regions from forested regions. Man-made structures such as oil platforms, roads and buildings are clearly evident, even though they are smaller than the 40-metre resolution of the Shuttle Imaging Radar. This is because of the strong backscatter of both metallic structures and corner reflectors formed where the walls of buildings join the surface of the ground.

A prominent part of the OSTA-1 payload is the Shuttle Imaging Radar antenna, in the left foreground in this view from the aft flight deck. Each instrument is covered by a white blanket to keep it within its range of operating temperatures.

A second pallet-mounted experiment was the multi-spectral infra-red radiometer measuring the intensity of solar radiation from the Earth's surface at ten different wavelengths. Five channels made observations between 2.1 and 2.4 micrometres, where minerals have a wealth of absorption features in their spectra. Thus carbonate minerals such as chalk could be distinguished from hydroxyl-bearing minerals such as clays. A third experiment consisted of a two-channel radiometer in the infra-red (at about 4.7 micrometres) to estimate the amount of carbon monoxide in the atmosphere at altitudes of approximately 7.5 and 11 kilometres. It was found that the density of carbon monoxide varies considerably, with both latitude and longitude, about a mean value of 120 parts per billion by volume.

Another remote-sensing experiment was an eight-channel, multi-spectral imaging sensor, operating at wavelengths between 0.4 and 0.8 micrometres, which is mainly in the visible part of the spectrum. The sensor detects variations in the colour of ocean surface waters by measuring the amount of solar radiation reflected from the surfaces of the oceans. These variations are related to the presence of chlorophyll in phytoplankton (the basis of the Earth's food-chain), to patches of plankton where cold, nutrient-rich water from the bottom of the ocean wells up, and to fish shoals.

An OSTA-1 Synthetic Aperture Radar image of the Californian coastline near Los Angeles clearly shows ships and a harbour breakwater.

An OSTA-1 Synthetic Aperture Radar image of northern Peloponnesia and southern Greece, covering an area approximately 50 × 100 kilometres. The straight canal of Corinth and the city of Corinth are clearly visible.

The fifth experiment consisted of a camera plus two imaging devices measuring surface reflectivity at wavelengths of 0.65 and 0.85 micrometres. Green vegetation has a low reflectivity at 0.65 micrometres due to the chlorophyll absorption band. A microprocessor was set to distinguish between water, barren ground, vegetation, or clouds/snow/ice.

Two further experiments were carried out in the crew's compartment. One was to take a film of lightning discharges by day and night using a hand-held camera. The short time available meant that only a very limited amount of data was obtained. The other, a precursor to Spacelab studies, was an attempt to grow dwarf sunflowers (*Helianthus annuus*) in space, and to determine the optimal amount of soil moisture. The seeds did not germinate in the period of abbreviated mission.

For the launch of STS-3, the huge external tank was dark coloured, unpainted, thus saving 250 kilograms. Both the propulsion and guidance systems worked well to put the Shuttle into the desired circular orbit at an altitude of 216 kilometres. A series of thermal tests was conducted with the Orbiter in different attitudes such as its nose facing the Sun, tail to Sun, cargo bay to Sun, or cargo bay to space; these tests raised no problems. The Canadian-developed Remote Manipulator System was controlled to hold, remove, move to different positions and stow again a 160-kilogram payload, the Plasma Diagnostics Package. Because heavy rains in California had flooded the lake bed runway several weeks before, use of the landing site at White Sands, New Mexico, was planned. High winds and blowing sand at this landing strip forced a one-day extension to the flight.

The OSS-1 payload on its pallet is being checked prior to its installation in Columbia's cargo bay.

Nine of the nineteen experiments conducted on the STS-3 flight are mentioned here. First, in order to help the design of a temperature control system for one or more experimental instruments, heat pipes were used to control the temperature of a canister between 5 and 25 degrees Celsius for a range of internal heat loads, and for external temperatures ranging from −75 to 100 degrees Celsius.

The Plasma Diagnostics Package measured parameters of the neutral and charged particle environment in various places around the Orbiter, at the end of the remote manipulator arm. The plasma density and composition varied from ambient values to a rarefied mixture with water vapour ions produced by the Orbiter. Disturbances were noted, both upstream and in the wake of the Orbiter, and when the Reaction Control System thruster or electron gun was operated. The difference in potential between the plasma and the Orbiter's electrical earth (or ground) was found to vary between −5 and +5 volts. Part of this can be explained by electrons and ions both arriving at and leaving the Orbiter's surface, and part by the motion of the Orbiter, covered by electrically insulating tiles, through the geomagnetic field. Larger deviations could affect electronic circuits and components on board.

Photographs taken at night from the aft flight deck window show that the Orbiter glows. It has been deduced that this glow, which is brightest at low altitudes, arises from an interaction between the Space Shuttle and the ambient atmosphere. The glow is greatest in the direction of the Orbiter's motion, known as the ram direction. It is probably produced by hydroxyl radicals formed when excited oxygen atoms, which have gained energy due to the Shuttle's motion, react with water absorbed on the Shuttle surfaces. These results give some cause for concern to experimenters using sensitive optical or infra-red instruments aboard the Space Shuttle.

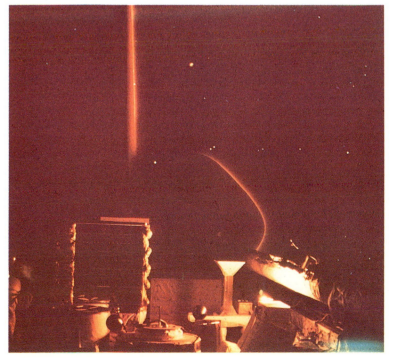

A night-time photograph of the OSS-1 payload reveals the Shuttle glow. As an electron gun is fired to investigate electrical charging of the Orbiter, a greenish glow occurs (lower right). The faint airglow originating from the Earth's atmosphere is evident at the top of the picture.

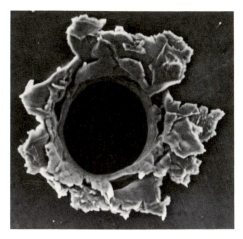

Scanning electron micrograph of a hole made by a micrometeorite in aluminium foil aboard STS-3; the diameter of the hole is 7.5 micrometres.

Another experiment involved the germination of seedlings of pine, oat and bean in orbit. These seedlings, grown in conditions of near zero-$g$, were to be compared with identical seedlings grown on Earth. As they grow, all plants produce lignin, a polymer which acts as a skeleton to maintain growth against the downward gravitational force. Since as much as 30% of the woody tissue of a plant may be lignin, this substance interferes with the extraction of wood fibres of a plant for paper and chemical cellulose.

The first British experiment to be carried aloft by STS consisted of 5-micrometre-thick aluminium foil sheets bonded to a plastic substrate. High-velocity impact craters are formed on the surface of the foil by micrometeorites in orbit around the Sun, with masses of about $10^{-7}$ micrograms (one microgram $= 10^{-6}$gram). Heavier micrometeorites penetrate the foil to a depth which depends on their velocity. Analysis of the foil therefore provides information on the number densities of micrometeorites around Earth. Such small particles are burnt up in the upper atmosphere, and cannot reach the Earth's surface.

The intensity of ultra-violet radiation from the Sun was measured for 20 hours at a time near solar maximum. The instrument was checked during the flight and recalibrated afterwards to obtain an accurate value of the solar flux in the ultra-violet part of the spectrum. It will be reflown on later Space Shuttle flights to investigate the variation of the flux during the well-known 11-year cycle of solar activity. The ultra-violet radiation is absorbed in the stratosphere where and when ozone is produced. The amount of atmospheric ozone was also found around the globe from the intensity of backscattered ultra-violet radiation. X-rays are emitted from solar flares, dramatic explosions on the Sun's surface. In order to advance understanding of the plasma physical processes that generate solar flares, three large analysers were carried on STS-3 to make the most thorough study of seven solar flares ever undertaken.

A student's experiment was also carried into space on STS-3. This compared the effects of weightlessness on the flights of insects with different sized wings. The feasibility of making identical-diameter latex spheres in space was also studied. This could have applications in medical and industrial research.

## Results from STS-4 to STS-8

The orbital test flights ended with STS-4 and the Space Shuttle became operational with STS-5. Missions STS-6 and STS-8 were of particular concern to Spacelab since they involved the Tracking and Data Relay Satellites (TDRS) to be used for the radio transmission (telemetry) of real-time data from the Spacelab experiments to the ground.

The STS-4 mission was almost flawless. A US Department of Defense satellite was successfully launched. The crew also carried out an engineering test of the Continuous Flow Electrophoresis System. Arranged jointly by NASA and McDonnell Douglas, this was the first commercial experiment to be carried out aboard the Space Shuttle. Biological cells were separated according to the electrical charges on their surfaces. The separation was preserved during descent by placing the samples in a freezer. Such experiments could lead to the manufacture of special drugs in space.

With STS-5 and its payload of 16 000 kilograms, came the first launching of commercial communications satellites. Once in space, on the sixth orbit of the Earth, at 300 kilometres altitude, the SBS-3 satellite was rotated on a spin-table to about 50 revolutions per minute, and then spring-ejected at one metre per second. After 45 minutes, the Orbiter was far enough away that the solid-fuel rocket, developed by McDonnell Douglas and termed the Payload Assist Module (PAM), could be ignited for 83 seconds. This launched the SBS-3 satellite towards its geostationary orbit. Twenty-four hours later, the procedure was repeated with Canada's Anik C-3 communications satellite. A few days later, each satellite was injected into a circular geostationary orbit by its own apogee kick motor. STS-5 was also the flight of the first two Mission Specialists, professional scientist–astronauts, as well as the Commander and Pilot.

The 'spacewalk', or Extra-Vehicular Activity (EVA), planned to last 3.5 hours, had to be cancelled. This was because a fan circulating air failed in one astronaut's spacesuit and there was a leak in the other. This new type of suit is lighter, more flexible and more durable than earlier ones which were tailor-made for each individual. Unisex suits are now made in three standard sizes.

Some experiments proposed by American students, the so-called Student Involvement Programme, were performed in the Orbiter's aft flight deck during STS-5. Also, seven self-contained canisters of experiments, mainly in the field of Material Sciences, were carried. Such canisters are termed Get Away Specials (GAS).

First used in April 1983, Challenger is 1100 kilograms lighter than Columbia. It has an external tank which is almost 5000 kilograms lighter than earlier models, making it possible to carry larger payloads than before. Challenger launched on STS-6 with its payload of three GAS canisters and the first Tracking and Data Relay Satellite (TDRS) into a circular orbit at 285 kilometres altitude, and 28.5 degrees inclination. Unfortunately, the McDonnell Douglas Inertial Upper Stage rocket did not fire for its full period, so that TDRS-A was placed in an eccentric orbit with an apogee of 35 400 kilometres and a perigee of 21 900 kilometres. The required geostationary orbit was circular at an altitude of 35 800 kilometres. The satellite was 'rescued', adjusting its orbit by making successive burns of its twenty-four attitude control motors (using hydrazine). Each manoeuvre was performed at apogee; after using 370 kilograms of propellant, the US$100-million TDRS-A reached the required orbit on 29 June 1983. After a later check-out using the White Sands Ground Station, the unmanned Landsat satellite and the STS-8 Orbiter, TDRS-A was declared to be fully operational. The satellite was at longitude 65 degrees West on attaining its geostationary orbit. It was then 'drifted' to its operational position of longitude 41 degrees West which it reached in mid-October 1983. As a result of the difficulties encountered with launching the TDRS-A, the launch of the TDRS-B was postponed. Thus, only one TDRS was available for the first Spacelab flight in November and December 1983, even though the flight plan called for two satellites. This necessitated changes 'at the eleventh hour'.

During the STS-6 flight, the two Mission Specialists performed EVA for 3.5 hours. Whilst carrying out several exercises in the payload bay, their new spacesuits functioned extremely well. Finally, during re-entry, various tests were carried out on the aerodynamic response of Challenger to evaluate the manoeuvrability of the new Orbiter.

Challenger arrives at Kennedy Space Center for the sixth Space Shuttle mission.

One of the rear Reaction Control System thrusters is fired during the STS-6 flight.

STS-6 astronauts Story Musgrave (left) and Donald Peterson explore the cargo bay that housed the TDRS-A satellite.

A unique picture of the Orbiter Challenger, taken from the Shuttle Pallet Satellite at a distance of some 300 metres.

Sally Ride, America's first woman in space, talks to ground controllers from the flight deck during STS-7.

The workhorse rests after 5 days in orbit for the
STS-5 mission. Columbia is towed slowly from
the runway where it landed at Edwards Air
Force Base in California.

The STS-7 flight of the Orbiter Challenger was particularly noteworthy as it marked the first flight of a woman astronaut in the US Space Program – Sally Ride. Also carried was the German satellite SPAS, the Shuttle Pallet Satellite. In this mission Challenger demonstrated the Shuttle's ability to rendezvous with and to retrieve another spacecraft. On the fifth day of the flight, Sally Ride used the Canadian-built remote manipulator arm of the RMS to deploy the SPAS. It was released and flew with the Orbiter at a separation of about 300 metres. In addition to a number of scientific experiments, SPAS carried a camera to obtain the first pictures of the Orbiter in space. After flying in formation for about one hour the SPAS was recovered, again using the manipulator arm. Later, the SPAS was released again and Challenger moved around it so that pictures could be taken of the Orbiter's reaction control motors being fired.

TDRS-B should have been launched during the STS-8 mission. However, since the Inertial Upper Stage rocket had malfunctioned on STS-6, the launch of TDRS-B was postponed until this problem had been solved. The launch of the TDRS-B satellite was rescheduled for late 1984. The Indian communications satellite INSAT-1B was launched during Challenger's second day in orbit at an altitude of 278 kilometres. A PAM motor was again used to put the satellite into a geostationary transfer orbit. Other activities included experiments to investigate the efficiency of heat pipes and experiments under zero-$g$ in the 12 GAS canisters carried on-board. Eight of the GAS canisters contained specially stamped postal covers. The formation of snow, contamination, exposure levels for ultra-violet film, and the influence of high-energy charged particles on integrated circuit memories were studied in the other four canisters. Further, the McDonnell Douglas Continuous Flow Electrophoresis System was operated again, the aim being to separate insulin-producing cells from other cells in a preparation of canine pancreas. This work could lead to a method of obtaining cells to be transplanted into patients suffering from diabetes.

The spectacular take-off at night was followed by a night-time re-entry and landing. Because the weather is normally good in Florida at night, the time available for the return to Earth is more than doubled. The STS-8 flight also provided a final check on the performance of the only Tracking and Data Relay Satellite. The Ku-band system antenna was deployed from the Orbiter payload bay, and data were transmitted to the ground via the geostationary TDRS-A. Also, the S-band system was used to transmit relatively high rates of data. The TDRS-A came through the test satisfactorily, transmitting excellent pictures and other data from aboard the Shuttle. By the Autumn of 1983, all was set for STS-9 and the first flight of Spacelab.

The spectacular launch of STS-8 was the first to
take place at night.

## Astronomy

Astronomy is the study of the Universe. It is investigated experimentally by analysing electromagnetic radiation emitted by celestial bodies. All wavelengths of the electromagnetic spectrum can be received by sensitive instruments operating in space, whereas observations from the ground can only be made in the visible region of the spectrum and in the radio region, which are the 'windows' through which the Earth's atmosphere transmits radiation.

Most stars are basically similar to the Sun (whose radius is $7 \times 10^5$ kilometres). They shine just as brightly, but appear as point sources of light because they are so far away. The nearest star to Earth is Proxima Centauri, some 4.2 light years distant. (A light year is the distance travelled by light through space in one year, that is, $9.5 \times 10^{12}$ kilometres.)

A typical galaxy, containing some hundred thousand million ($10^{11}$) stars, is shaped like a disc and often exhibits spiral structure, like a catherine wheel. It is about 100 000 light years in diameter. Between the stars are interstellar dust and interstellar gas (mainly hydrogen) which, if dense enough, is a region of star formation. Interstellar gas clouds are sometimes visible as nebulae because they are illuminated by bright hot new stars within them.

# SCIENCE AND TECHNOLOGY FROM SPACELAB

*'For all knowledge and wonder (which is the seed of knowledge) is an impression of pleasure in itself.'*

Francis Bacon

Stars in the Orion Nebula illuminate clouds of interstellar hydrogen, giving rise to the red colour evident here. The famed Horsehead Nebula, part of a large dust cloud, absorbs the radiation and so appears dark.

(Above) This galaxy in Ursa Major (M 101) is a typical spiral galaxy. It contains many billions of stars and is 14 million light years from Earth.

(Below) The Space Telescope is depicted as it will appear in orbit. Electrical power is provided by the arrays of solar cells, which are deployed in orbit, and the information on distant objects sent to Earth by radio. Using this, astronomers will be able to 'see' objects near the edge of the Universe.

The best ground-based telescopes can see some $10^{10}$ galaxies; the most distant of these is about $10^{10}$ light years (about $10^{23}$ kilometres) away. All these galaxies are moving away from all the others, including the Milky Way, our own Galaxy which contains the Sun. The entire Universe is expanding. The receding motion of distant galaxies causes their emitted radiation to be shifted towards the longer wavelengths, that is, to lower frequencies. (Such a Doppler shift also accounts for the increasing frequency of a train whistle heard when the train is approaching, and the decreasing frequency heard when the train is receding.) The 'red shift' towards longer wavelengths is more pronounced at larger velocities and is, thus, greater the more distant the receding cosmic source. Fainter, and even more distant, galaxies will be seen from the far ultra-violet to the near infra-red when the 2.4-metre-diameter Space Telescope, to be launched by the Space Shuttle in 1986, is in Earth orbit. Towards the edge of the visible Universe are quasars, quasi-stellar objects. These may be extremely energetic and bright galaxies, receding at nearly the speed of light. They were formed soon after the 'Big Bang', the origin of the Universe itself and the cause of its expansion.

A star is born from the interstellar gas, due to self-gravitational attraction. When it is sufficiently compressed, and thus heated, nuclear reactions occur in the star's central core. In this process of thermonuclear fusion, hydrogen is converted into helium and the energy so produced is radiated from the star's surface as electro-magnetic radiation (including light).

Most stars follow evolutionary patterns which are similar. Stars live by burning their thermonuclear fuel and then die when all this fuel has been exhausted. The manner of a star's death depends upon its mass. If the star's mass is not too large (up to about one solar mass), it becomes a 'white dwarf' star, whose radius is approximately

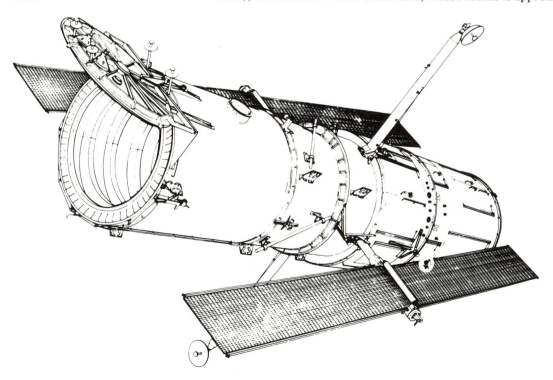

This diagram illustrates a black hole at the centre of a galaxy. Matter spirals in to the accretion disc, where X-radiation and cosmic rays are produced.

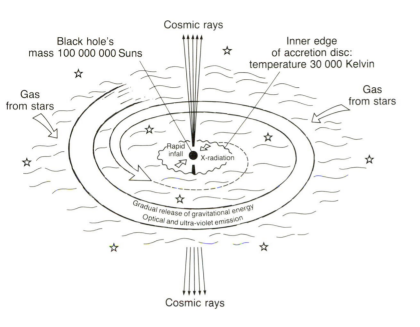

the same as that of the Earth. This star then fades and cools, eventually dying as a cold black object. If the star's mass is somewhat greater than that of the Sun, however, it explodes catastrophically, causing a 'supernova'. At the centre remains a neutron star formed from nuclear particles – its radius is only a few kilometres. If this star contracts further, it eventually collapses forming a 'black hole'. The gravitational attraction of a black hole is so intense that neither matter nor radiation can escape from it. Hence, a black hole cannot be observed directly, but its effects – due to gravitational attraction on neighbouring gas or matter from a companion star – can be studied. Matter attracted into such an accretion disc disappears into the black hole, never to be seen again. In this energetic accretion process, X-radiation is emitted; the observation of X-radiation from an accretion disc therefore indicates the presence of a black hole.

The violent supernova of 1054 produced the Crab Nebula, which has a neutron star at its centre which rotates very rapidly, thirty times per second, and is termed a pulsar. The expanding Crab Nebula, at a distance of 6000 light years and now 5 light years across, produces most unusual radio, optical and X-radiation emissions due to the motion of extremely high-energy charged particles in its magnetic field.

The most energetic electromagnetic radiation observed in the Universe, gamma-radiation, is produced by the interaction of cosmic rays – electrons and protons accelerated to $10^9$ volts or more – with the interstellar gas in the disc of a galaxy. Only a few sources of gamma-radiation have been identified.

From their vantage points in space, instruments mounted on Spacelab's pallets or outside the airlock can enable observation outwards from the Earth. Telescopes or cameras sensitive to particular regions of the electromagnetic spectrum receive radiation emitted by the many different types of astronomical objects that populate the Universe. Included amongst these are new stars, old stars, regions in which stars are being formed, galaxies and clusters of galaxies. Heavy, large-aperture telescopes incorporating the latest design features and carried into space aboard the Space Shuttle will enable astronomers to make observations of regions near the outer edge of the observable Universe.

## Solar Physics

The Sun is a typical star, of diameter $1.4 \times 10^6$ kilometres, or 109 times the Earth's diameter. The centre of the Sun consists of hydrogen, at a temperature of over 10 million Kelvin, which is converted into helium at a rate of $5 \times 10^9$ kilograms per second. In this hydrogen-burning phase of its life-cycle, the Sun is now middle-aged. The mass of the Sun is so large, however, that it should continue to burn steadily for another $4 \times 10^9$ years or more. This energy is radiated into space from the photosphere, the apparent surface of the sphere of gases called the Sun. The temperature at the photosphere is about 6000 Kelvin, and most of the radiation is emitted in the visible part of the spectrum. The chromosphere is the region some 4000 kilometres above the photosphere where helium was discovered by its spectrum and where ultra-violet radiation is emitted. Above this layer the gases become far more rarefied but the temperature increases sharply; the solar corona is at a temperature of over one million Kelvin and emits X-radiation strongly.

A coronagraph aboard the Solar Maximum Mission satellite produces an artificial eclipse of the Sun. The coronagraph observes brightness variations in the solar corona. Interpreted as different densities, these are shown here (for one quadrant) as different colours.

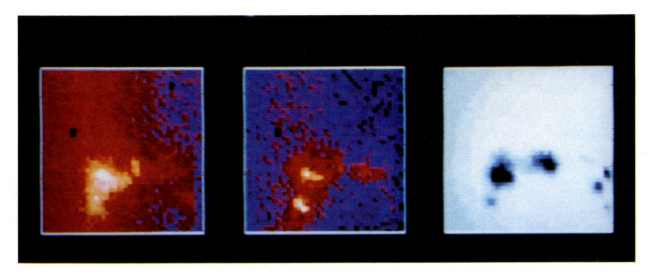

(Above) The repaired Solar Maximum Mission satellite observed a dramatic and energetic solar flare on 25 April 1984. The left-hand and central images were observed using X-rays, while the right-hand image, showing the sunspots directly, is in the visible part of the spectrum. Each image is of an area 7 arc seconds × 7 arc seconds on the Sun.

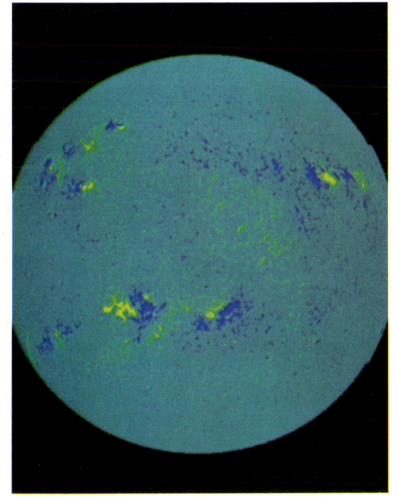

(Left) Neighbouring sunspot pairs have opposite magnetic polarity, as shown by the blue and yellow colours, and are connected by magnetic field lines which loop up through the corona. In the Sun's northern hemisphere the polarity of the 'leading' sunspot (in the direction of solar rotation) is opposite to that of the leading sunspot in the Sun's southern hemisphere. Basically, this is because the Sun rotates more rapidly at the equator than at the poles. The polarity of the leading sunspot reverses each 11 years.

The paths of magnetic field lines through the corona are revealed by brightness variations observed during a solar eclipse.

In this high-resolution photograph of the Sun's photosphere, the plages are bright and sunspots dark. The speckled texture illustrates the granules.

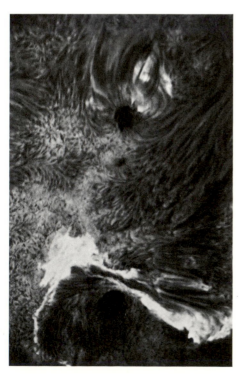

The Sun is the only star that can be studied in detail, being relatively close to the Earth; it is thus a most valuable astrophysical laboratory. Details of the photosphere that are about 1000 kilometres across can be distinguished by a good telescope that has a resolution of one arc second. (Such a telescope can distinguish one edge of an object of diameter one centimetre from the opposite edge at a distance of 2 kilometres.) On this scale of 1000 kilometres the Sun is not glowing uniformly, but reveals brightness variations termed granulations. At the centre of each granule, matter rises and, at the edge, it falls back into the interior of the Sun. Each granule is a convection cell; it is somewhat similar to a cumulus cloud in the Earth's troposphere.

Strong magnetic fields are generated by electric currents flowing in dynamos below the photosphere. The outward-convecting solar plasma twists and tangles the magnetic field lines whose inherent buoyancy takes them out to the photospheric surface. Where the magnetic field lines are squeezed together, strong magnetic fields are produced: these are termed magnetic active regions. They can be hot and appear bright, in which case they are known as 'plages' (the French word for beaches). Alternatively they can be cool and appear dark as 'sunspots', blemishes on the bright solar disc. The magnetic field lines form arches reaching into the corona and linking nearby sunspots of opposite magnetic polarity. The arched magnetic field structure may explode, shooting a gigantic cloud of hot plasma through the corona, out into interplanetary space.

By observing the positions of sunspots from day to day, Galileo Galilei discovered that the Sun rotates once in 27 days at low solar latitudes, but in a time which exceeds 30 days at high solar latitudes. The Sun does not, therefore, rotate as a solid body. Differential rotation causes solar magnetic field lines produced by dynamo action to be distorted and twisted on a large scale. Such processes are believed to explain the 11-year cycle of solar activity, whereby the number of sunspots waxes and wanes each 11 years. At the maximum of solar activity, for example during 1958, 1969 or 1980, sunspots occur near the Sun's equator. A few years after solar maximum, sunspots begin to appear at medium solar latitudes and to migrate towards the solar equator over the next 3 years. This is shown on a plot of sunspot positions observed over many decades which, for obvious reasons, is known as a butterfly diagram.

Energy stored by the solar magnetic field can be violently converted into other forms in a solar flare. The X-radiation output of the region can increase a hundred-fold, and the ultra-violet flux ten-fold, within a minute. Electrons and protons are accelerated to energies of one million volts in this explosion. When energetic electrons collide with the photospheric gas at the feet of the arched structures they produce bursts of X-radiation. This solar outburst usually decays away in an hour or so. The electrons produce intense radio emission, as synchrotron radiation, in the strong magnetic field.

Observations made aboard the Solar Maximum Mission satellite in 1980, as each large, dark sunspot group appeared on the solar disc and evolved, showed that the output of solar radiation decreased, typically by 0.1%. Thus the 'solar constant', the amount of energy received above the Earth's atmosphere, about 1.38 kilowatts per square metre, is not strictly speaking a constant! Such day-to-day variability of the incident solar energy flux could account, in part, for the variability of the Earth's weather. Long-term changes of the solar output, on time-scales of decades, centuries or millennia, could help to

(Below) The positions of sunspots have been carefully recorded over the last hundred years, as shown in the upper part of this figure. The proportion of the Sun's disc that is covered by sunspots varies, by up to 0.4%, over the 11-year solar cycle, as illustrated in the lower part.

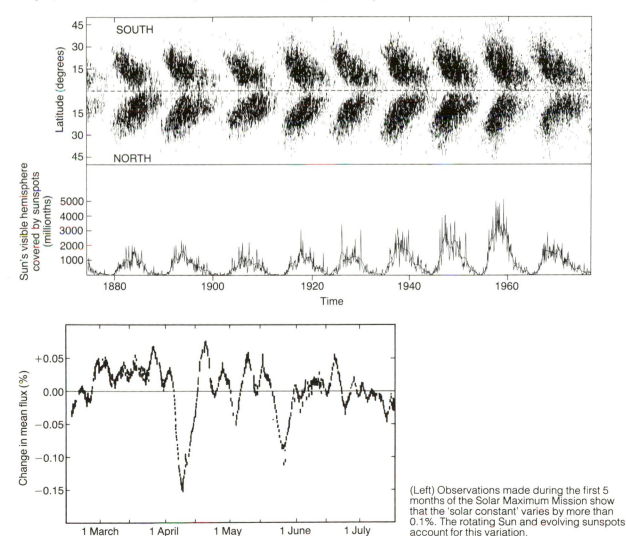

(Left) Observations made during the first 5 months of the Solar Maximum Mission show that the 'solar constant' varies by more than 0.1%. The rotating Sun and evolving sunspots account for this variation.

explain climatic changes. As yet, there are no direct observations of the solar constant over such long periods. There are some indications that the flux of solar ultra-violet radiation changes by a considerable fraction of 1% as the Sun rotates once each month. Such changes are likely to cause at least part of the natural variability of the amount of stratospheric ozone.

Specialised instruments aboard the Space Shuttle or Spacelab can make detailed measurements of radiation from the Sun, the source of energy for all life on Earth. Such measurements are, therefore, important not only to astrophysicists but to all inhabitants of the planet Earth. The particular advantage gained by using the Space Shuttle is that repeated measurements can be made with identical equipment simultaneously, or at different times, at different phases of the solar cycle.

## The Earth in space

The environment of the Earth is strongly influenced by the Sun. The solar corona, accelerated away from the Sun, becomes supersonic. Moving at typically 400 kilometres per second, a certain volume of plasma then takes 4 days to reach the Earth. In the interplanetary medium near the Earth's orbit, the density of this solar wind plasma is usually about 5 million electrons per cubic metre, with the same number of ions – mainly protons (hydrogen nuclei) – per cubic metre. After an explosion on the Sun, plasma of higher velocity – and density – compresses the ambient solar wind plasma, giving rise to a shock front moving through the interplanetary medium, often at around 800 kilometres per second. This radially outward moving solar wind plasma takes solar magnetic field lines with it. Although one end of each particular field line is rooted to the photosphere, the other is dragged out into the interplanetary medium. As a consequence of the Sun's rotation, all interplanetary magnetic field lines have a spiral shape.

This diagram illustrates the interaction between the solar wind and the Earth's magnetic field, which forms the three-dimensional magnetosphere. Its outer boundary, the magnetopause, is shown in yellow. Upstream of the magnetopause is the bow shock, coloured brown. Important features within the cut-away section of the magnetosphere are also shown. This representation is based upon observations made from spacecraft.

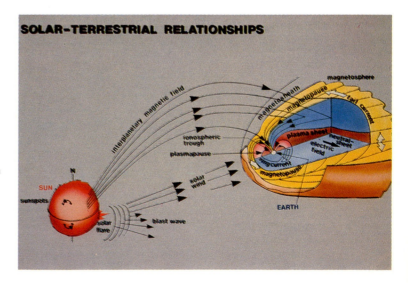

The Earth possesses a magnetic field. This is produced, by dynamo action, in the Earth's liquid core. The geomagnetic field has a shape resembling that produced by an intensely strong bar magnet, set near the centre of the Earth, and which is aligned some 11 degrees away from the rotational axis. At the Earth's surface the magnetic field is about $5 \times 10^{-5}$ teslas. Further away from the Earth, the geomagnetic field strength falls off as the inverse cube of the geocentric distance. At a distance of 10 Earth radii towards the Sun, this field has decreased to about 20 nanoteslas, which is about four times greater than the usual strength of the interplanetary magnetic field. Here an important boundary is formed.

The auroral oval is both brightest and widest near midnight, as shown in this view taken from a high-altitude satellite; the sunlit atmosphere is seen upper left. The Earth's geometry has been added using a computer.

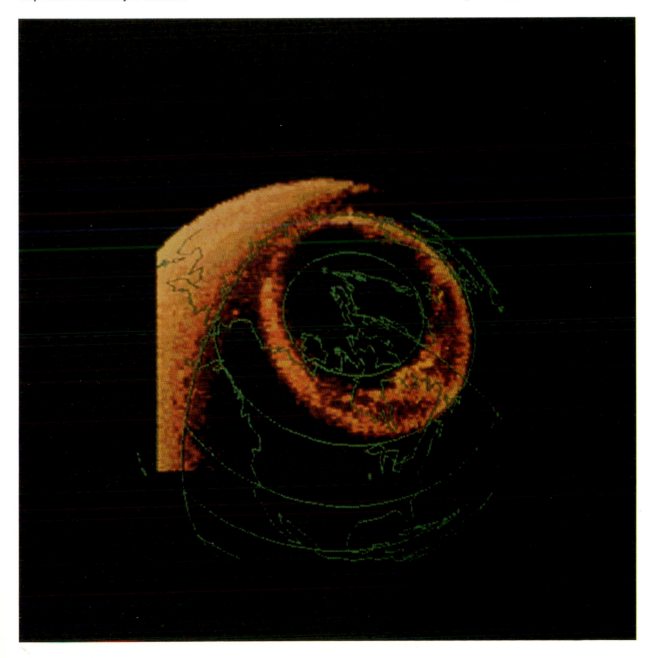

Called the magnetopause, this is a boundary between the onrushing solar wind plasma (on the outside) and the compressed geomagnetic field, the magnetosphere (on the inside). The solar wind is deflected around the edge of the magnetosphere, which acts as an obstacle to its flow. Because the solar wind flow is supersonic, a shock front is produced upstream of the magnetosphere. This arises in just the same way as a bow shock is formed ahead of a model of an aeroplane in a supersonic wind tunnel. Such a shock wave accounts for the sonic bang produced ahead of Concorde when it is travelling at supersonic speed.

The solar wind plasma draws out into a long tail the magnetic field lines fixed to the Earth's surface on the dark side of the Earth. This resembles the tail of a comet; it reaches beyond the Moon's orbit, about 60 Earth radii distant. Some plasma crosses the magnetopause, arriving inside the magnetosphere to mix with that which has moved up from the polar ionosphere. In the centre of the stretched tail of the magnetosphere, complex plasma physical processes occur which result in the acceleration of electrons and ions to a few thousand volts. These atomic particles are catapulted Earthwards and hit the top of the Earth's atmosphere. At a height of 100 kilometres or so, they give rise to the Aurora Borealis observed in the north and Aurora Australis seen in the south, at geomagnetic latitudes between 65 and 70 degrees. Thus, under normal conditions, they are observed on every cloudfree night over northern Scandinavia and Alaska, or from Antarctica. These eerie and dramatic, rapidly moving lights in the night sky have colours that depend upon the species of atmospheric gas being bombarded by the electrons.

When a faster than normal solar wind stream hits the magnetosphere, it causes a geomagnetic storm. Then brighter auroral lights are probable at latitudes some 5 degrees – or even 10 degrees – lower than usual. At such times of increased solar and geomagnetic activity, aurorae will be seen over Scotland, southern Scandinavia, northern parts of the Federal Republic of Germany, or near the border between Canada and the USA.

Because the solar wind is continually fluctuating in strength, the magnetosphere – a very dynamic and responsive system – is never steady or dull. Some electrons and ions, which avoid the fate of plummeting directly into the atmosphere in the auroral zones, are accelerated to form the van Allen radiation belts. They may eventually be lost by processes involving interactions with whistler-mode electromagnetic waves of audio frequency, or by other plasma instabilities.

Instruments aboard Spacelab in a high-inclination orbit traverse the region of the Earth's plasma environment where these processes are occurring. Other instruments can measure properties of the region around Spacelab, and investigate changes to the space environment produced by the Space Shuttle itself. Such experiments sense the medium in a 'passive' manner. In contrast, beams of electrically charged particles and plasma beams can be ejected into space from Spacelab. Such experiments in Space Plasma Physics are examples of 'active experiments' in which space is used as a vast natural laboratory; they produce scientifically interesting changes in the space environment which can be measured by the instruments onboard.

# Observations of the Earth's atmosphere and surface

The Earth's atmosphere allows part of the Sun's radiant energy that is incident upon it to reach ground level. That is, it transmits solar radiation, at least partially. The remainder it either reflects back into space or absorbs. The absorber can be airborne dust, a cloud, or molecular gases such as water vapour, ozone or carbon dioxide. These absorbers, and the Earth's land and sea surfaces, emit this absorbed energy at longer wavelengths, as infra-red radiation. Some of that energy is radiated into space and some is radiated back towards the Earth's surface. The fate of solar and terrestrial radiation determines the radiation budget of the atmosphere. The atmosphere – and oceans – transport heat from the equatorial to the polar regions. Together with the Earth's rotation and the presence of the sea, land and mountains, these factors account for the Earth's weather and climate.

These concepts can also be applied to understanding observations, made from space, of radiation that has passed through the atmosphere. For example, beams of infra-red radiation from the Sun can be viewed at sunrise, by a radiometer (an instrument measuring energy flux) aboard Spacelab. Such observations provide information on the infra-red absorbing gases at different heights in the atmosphere. Alternatively, a spaceborne instrument can look vertically down on the Earth. It may receive infra-red, or microwave, radiation at different wavelengths emanating from different atmospheric layers or from different types of rock or vegetation on the ground. Further, photographs taken in the visible region of the spectrum clearly show

The fate of solar radiation incident upon the Earth's atmosphere, shown here averaged over all latitudes, longitudes and seasons, plays a crucial role in determining the Earth's climate. The 51% of solar radiation which reaches the Earth's surface (left) is lost as shown on the right.

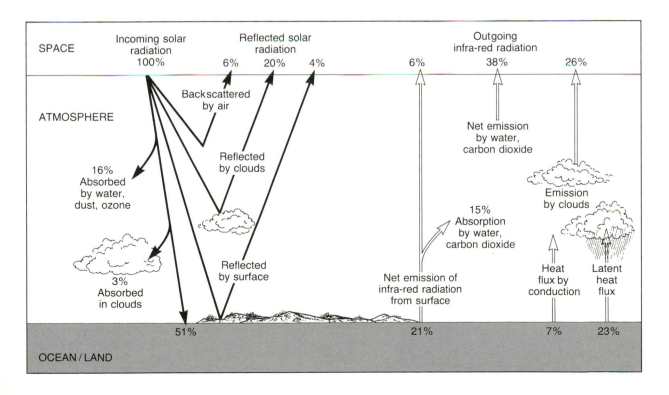

the nature of the Earth's surface and also the presence of clouds, by the sunlight that they reflect. Photographs taken in the near infra-red show a land surface that is bright (hot) by day and dark (cold) by night, a sea surface whose temperature is only a little higher in the daytime than at night, and cloud tops that are always relatively very cold.

All these are examples of 'passive remote sensing'. The Earth's surface or atmosphere is investigated remotely from space, by receiving radiation provided – in one form or another – by Nature. There is a complementary way of carrying out remote-sensing studies from space; this is the 'active' technique. Here a source of radiation at a chosen wavelength is operated on the satellite, and pulses of radiation are shone down upon the Earth and its atmosphere. These could be from a laser tuned to a line in the absorption spectrum of an important trace atmospheric species, or from a radar source.

A particular application of a radar system aboard a satellite is one designed to find the height of the satellite above the Earth's surface. A radar system on the satellite transmits radio pulses which are reflected by the Earth's surface. The echo is detected at the satellite, and its range – the height of the satellite – is derived from the time delay. This is done knowing that the radar signal travels at the speed of light; in fact, small corrections have to be applied accounting for propagation through the ionosphere and the atmosphere. Precise information on the height of the satellite enables the Earth's shape to be determined. This is related to the boundaries between the plates which, according to the theory of continental drift, move over the Earth's surface at a speed of a few millimetres per year. The radar also gives other information of interest in geophysics. Examples here are the height above sea level of the Antarctic and Greenland ice sheets, the heights of waves on the ocean, wind speeds over the ocean and the location and speed of oceanic currents.

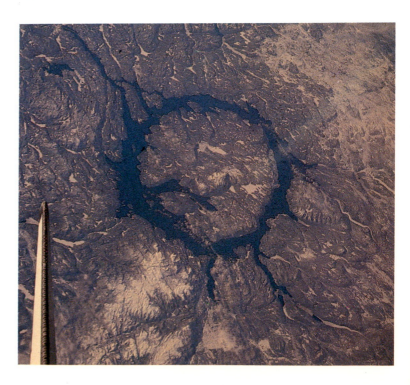

This is a fine example of the Earth's terrain photographed from space. It shows the Manicouagan area of Quebec, Canada, and illustrates the local geological devastation caused by a meteor impact that occurred long ago.

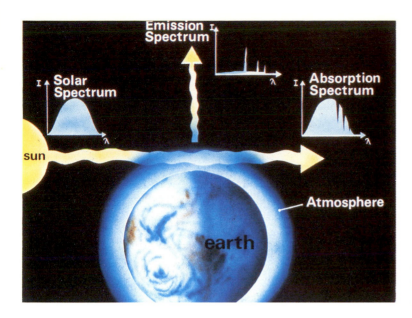

(Left) This diagram illustrates how the Earth's atmosphere can be studied in absorption (of solar radiation) or in emission. The relative amounts of different species of gas molecules determine the relative intensities (I) of features in the spectrum at different wavelengths ( λ ).

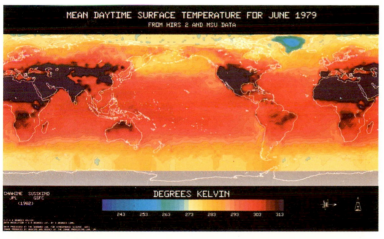

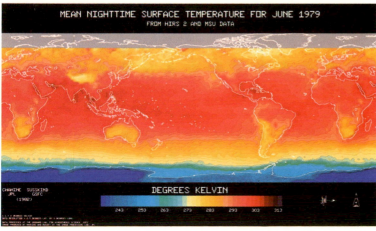

Temperatures at the Earth's surface can now be derived globally from infra-red and microwave measurements made aboard satellites. Higher temperatures are indicated in yellow and red, and the highest in brown, with temperatures below freezing being shown in green and blue. The temperatures shown are for June 1979, in the upper part for daytime conditions and in the lower part at night.

(Above) The Amazon basin at Coari, central Brazil, is viewed through thick cloud by a SAR instrument carried aboard STS-2. Hunting were joint official observers with RAE, Farnborough, to the SIR-A investigation team.

(Right) Photograph of Europe taken from a US National Oceanic and Atmospheric Administration (NOAA) satellite on 20 February 1984, in the visible region of the spectrum. The spiral of cloud is associated with the depression to the west of Ireland.

(Below) The weather map for 20 February 1984, showing the surface pressure (in millibars) and fronts, corresponding to the satellite photographs on the right.

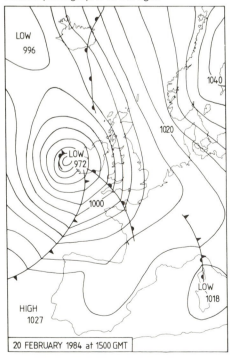

Long-wavelength microwaves and radio waves penetrate clouds, enabling the Earth's surface to be viewed through the clouds. All-weather, year-round, day-and-night coverage is important for systematic studies of the surface. Synthetic Aperture Radar (SAR) systems are sophisticated radars whose observations can be processed to give a two-dimensional representation like a photograph. These show ocean wave and current patterns, ice fields and the interaction between coastal and oceanic regions. Spacelab can readily house a large SAR instrument and supply the great amount of electrical power that it needs.

Remote-sensing satellites are often placed in orbits near 800 kilometres altitude. From there, their instruments (for example those aboard Landsat) have a resolution of some tens of metres over an area that is typically 200 kilometres by 200 kilometres. At a lower altitude, for example that used by the STS Orbiter, the resolution would be correspondingly improved, but the viewing area reduced. Geological features with an overall scale measured in kilometres are clearly delineated from space observations, and new mineral deposits found. New maps are being produced – for some areas of the world for the first time.

Photograph of Europe taken simultaneously from the same NOAA satellite in the infra-red. The whiter parts of the clouds are colder and, therefore, higher in the atmosphere.

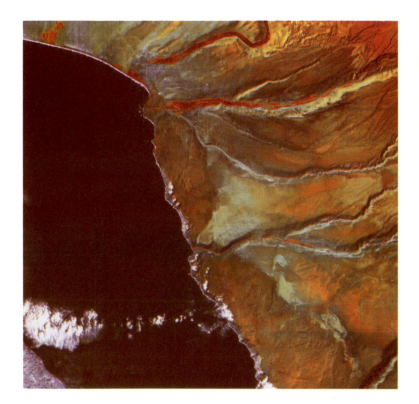

This false-colour image was obtained by the Modular Opto-electronic Multi-spectral Scanner (MOMS) on board STS-7 on 8 June 1983. It shows the border between Peru and Chile and the Pacific Ocean. The vegetation growing in the river valleys and certain rocks emit distinctive infra-red radiation and are shown here in red. Cumulus clouds are also evident, in white.

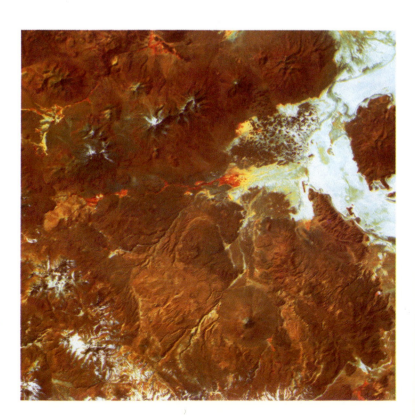

Another MOMS image shows the mountainous border between Bolivia and Chile. Volcanic peaks and lava flows are seen; the mountain peaks (at the bottom) are snow-capped. The white area (upper right) is a large salt pan, the Salar de Coipasa. It is interlaced with yellow-green streaks indicating the presence of salt water.

# Microgravity – Material Sciences

Weightless conditions prevail within a vehicle in orbit around the Earth. Inside the Spacelab module this would be exactly so – the so-called zero-$g$ environment – if the atmosphere exerted no drag, if all the astronauts were perfectly still, and if none of the Orbiter's thruster engines was being fired. But then the astronauts could not do any experiments! Thus the very presence of crew members moving around the Orbiter and the operation of thrusters controlling its orientation in space produce small accelerations on substances placed in the module. The desired weightless condition – or zero-$g$ environment – is changed into the almost weightless condition – microgravity environment – by the necessary presence of men aboard the Space Shuttle–Spacelab combination. A typical value for the acceleration inside the Spacelab module is $10^{-4}$ to $10^{-5}$ times the acceleration due to gravity at the Earth's surface, which is very attractive for Material Sciences research.

The microgravity environment within Spacelab permits the effects of small forces, which would be dominated by the one-$g$ force on Earth, to be evaluated. Experiments can be carried out which are of either a fundamental or an applied nature – they can even lead to the discovery of new processes. Certain experiments on the behaviour of liquids can be performed for the first time. The outcome of these experiments can be viewed in the module and – via the video link – simultaneously at the Payload Operations Control Center.

The phenomenon of convection in a fluid that is heated from below is evident in a pan of boiling soup or in a cumulus cloud in the one-$g$ environment at the Earth's surface. A liquid heated from below becomes less dense and, by buoyancy, rises. Elsewhere colder, more dense liquid sinks under the influence of gravity, and a convection cell becomes established. Many convection cells exist in a volume of liquid placed on a hot plate, and its upper surface exhibits the characteristic hexagonal pattern of a honeycomb. Such convection cells, driven by forces caused by differences of density, produce mixing of the liquid. They transfer mass and energy (as heat) within the liquid. However, this convection process cannot occur in a zero-$g$ laboratory.

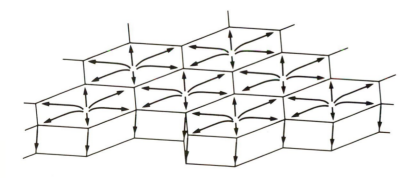

This diagram illustrates how convection cells appear in a fluid. Material rises at the centre of each hexagonal cell, flows outwards along the upper surface, and falls at the edges.

In Earth orbit, another convection process – Marangoni convection – can be investigated. Marangoni convection is driven by small differences in the forces due to surface tension at the upper surface of the liquid; these differences are caused by changes in temperature or density along the surface. This effect is swamped in the Earth-based laboratory by normal convection. In microgravity conditions the minute forces give rise to a flow at the surface from the region of low to high surface tension, from hotter to colder regions, setting up a circulatory flow within the bulk of the liquid.

A further phenomenon which can occur in the presence of a temperature gradient is thermal diffusion. In a gas, this causes the partial separation of atoms according to their mass. This is the Soret effect.

It is also interesting to study what happens at the interface between a solid and a liquid, or between a liquid and a gas, under microgravity conditions. At a liquid surface, surface tension forces act to cause the surface to assume a spherical shape, whereas on Earth, gravitational forces predominate to form a plane surface except at the edge. If there is a strong force of attraction, or adhesion, between the liquid and the solid surface which forms the container for the liquid, the liquid rises up the container wall, wetting it. The shape of the meniscus of a liquid in a beaker is thus far more curved in space than on Earth. Conversely, a perfectly plane liquid surface can only be formed in space if the beaker walls are curved correctly.

A third area of study is the combination of surface and convection phenomena under conditions of microgravity. Such fundamental physical processes need to be investigated before the benefits of space can be realised in technological terms, that is, by the manufacture of high-quality products. This is because any process, such as the growth of a crystal or the development of a special alloy, involves the behaviour of a liquid in microgravity and also the interface between liquid and solid materials. For example, a large crystal can be grown by applying heat to material from which the crystal is to be formed, melting it, placing a small near-perfect crystal in the liquid 'melt' (that is, the crystal material in the molten state) and allowing this to cool gently, in the absence of gravity-induced convection or turbulence. A pure crystal, free from defects, may also be grown from the vapour in the absence of convection. An alloy can only be made by taking appropriate proportions of the constituent metals, heating them so that they become molten and allowing the resultant mixture to solidify. A uniform alloy formed of metals of widely different densities (such as aluminium, $2.6 \times 10^3$ kilograms per cubic metre, and lead, $11.4 \times 10^3$ kilograms per cubic metre) can be more readily made in the microgravity conditions of space than in the one-$g$ environment of an Earth-based laboratory. On Earth the walls of a container act as nucleation centres in the solidification process. These effects can be avoided in space where a liquid does not need to be confined within a container. Also contamination is avoided.

New concepts of materials technology, drawing upon the wetting behaviour of melts with respect to solid surfaces, are becoming real possibilities. Furthermore, studies of the wetting effect of a lubricant placed between two metal surfaces will have practical relevance. In terms of space engineering, for example, bearings will be required to have a long life, operating reliably in the space environment. Thus fundamental studies of tribology, that is, of the frictional forces between two surfaces and the behaviour of lubricants, have first to be carried out in space.

Spacelab forms the ideal space laboratory for all such experiments. The Payload Specialist can use his skills while performing a wide variety of experiments, so making this a doubly valuable laboratory.

## Microgravity – Life Sciences

For the first 2 or 3 days in Earth orbit, one in two astronauts experiences unpleasant, though not debilitating, nausea and dizziness. This condition of 'space sickness' is similar to motion sickness experienced in cars or aircraft, or at sea. It is euphemistically called the 'space (mal)adaption syndrome'.

Space sickness occurs because the human body's balance mechanism does not always rapidly adapt to the microgravity environment. On the Earth's surface, almost everyone is able to maintain balance and orientation by the brain's ability to combine information received from various sensory organs such as the eye and the skin. The balance mechanism, or vestibular system, relies strongly on the otolith organ in the inner ear sensing accelerations. Under weightless conditions the otolith cells send no signals to the brain. In order to understand this important factor and just how the vestibular system interacts with other organs and senses, further experiments have to be performed using Man in space as the subject. Astronauts seem to become disoriented when the visual signals conflict with those from their inner ears and with information stored in their brains as a result of spending most of their lives in a one-$g$ environment. The increased knowledge will undoubtedly benefit those on Earth who suffer from disorders of the vestibular system.

The cardiovascular (blood circulation) system is also affected by the absence of gravity in space. The performance of the heart depends upon the distribution of blood to different parts of the body. In space, one or 2 litres of blood are redistributed from the lower part of the body to the head, neck and thorax. This explains the shrinking of the lower limbs, a situation known as 'chicken legs', and also facial swelling in space. Such changes in blood distribution could not only cause hormonal or metabolic changes but could also have effects on pituitary and adrenal secretions. There is a decrease in the volume of plasma, the colourless liquid component of blood; the red blood cell count also decreases. Further, urinary excretion increases. The reduced body weight becomes stable after a few days in orbit.

The locomotor system, the body's skeleton and muscles, is partially rested in space. Intervertebral discs stretch, explaining the temporary increase in the heights of astronauts, by as much as 5 centimetres. The phenomenon of atrophy – wasting away – occurs in the skeleton and muscles. Demineralisation of the bones takes place and there is a corresponding loss of calcium, through the urine, of about 0.5% of the total calcium content of the body per month.

Studies of these effects are important not only in relation to space medicine but also for the insight that they provide into human physiology. Besides Man, other animals, animal cells and plant cells are studied in space to obtain a better understanding of the role that gravity plays in the fundamental processes of life. Concerning plants,

Man-in-space experiences many unusual medical stresses. Here, a subject in the laboratory experiences the application of reduced pressure to his lower body, while the circulation of his blood is monitored.

the orientation of growing roots and stems depends on gravity. In space this sense of direction is absent.

Radiation in space can be a hazard for Man and other animals in Earth orbit. This is due to cosmic rays and other charged particles, particularly those trapped in the van Allen radiation belts. During and following a solar flare there is an additional radiation hazard. The flare, which is more likely near the maximum of the 11-year cycle of solar activity, gives rise to a burst of ultra-violet and X-radiation and, a day or two later, to enhanced fluxes of energetic charged particles, particularly in the polar regions. The walls of a space laboratory or Space Station afford protection from such radiation. However, a spacesuit donned for Extra-Vehicular Activity is not thick enough to prevent an astronaut's being exposed to dangerous levels of radiation from a solar flare which occurred before his EVA. Experiments have also been performed exposing bacteria and other cells to the harsh radiation environment of space.

The vacuum of space also presents a hazard to an astronaut. There should be no significant leak in the spacesuit or space vehicle which contains the oxygen – or air – that the astronaut breathes in order to stay alive. Thanks to Spacelab's shirt-sleeve environment, the astronaut only dons his spacesuit for work outside the module.

Spacelab is ideal for investigating the problems experienced by Man in space and for studying Man's basic physiology. This is mainly because large pieces of equipment can be installed and operated in this large space laboratory.

## Technology

Experiments on technology are normally outside the domain of both 'pure science', such as Astronomy, and 'applications', such as operational communications satellites. They can, however, help with the development of equipment and methods in both areas. Thus, for example, the development of a telescope or a sensor for an Astronomy experiment, or research leading to the use of a certain radio frequency in a communications satellite, would be classified as 'technology'. Other examples include the development of cryogenic cooling techniques, the use of microwaves for imaging the Earth's surface, the development of laser methods for measuring distances accurately and the lubrication of moving parts to be used in space. Further examples would be the testing of new construction methods in space, needed before a Space Station can be made, and the means of erecting large antennae. Both of these could involve manipulation by robots.

Spacelab is admirably suited for all these technological activities because of its capability to carry large masses and volumes into space, the fact that the equipment can be brought back to Earth and the presence of astronauts who can make judgements and decisions. Thus Spacelab's use as a test-bed in a Research and Development programme for a particular piece of space equipment is invaluable. This, coupled with its demonstrated ability to provide basic scientific data, can lead to the best possible use of equipment for the long times required for operational applications.

A particular case is that of an operational satellite for making long-term Earth Observations. Although Spacelab itself is not suitable for this task, it can be used for developing the related technology.

## The decision to fly

The first step in the definition of a Spacelab mission is the awareness that a flight is the best way to provide data for the investigation of a particular area of science or technology. Funding must be found not only for the actual flight but also for developing the instrumentation on board, and for the necessary ground operations. These funds may be provided through NASA or ESA, or indeed any such Agency throughout the world representing the interests of Astronomy, Earth Observations, Material Sciences, Life Sciences, etc. All funding in NASA-sponsored missions is normally provided by NASA. With ESA, the Agency pays all flight-related costs whereas the costs of developing the equipment for the experiments are usually paid by a body within the Member State from which the experiment originates. Further, it is hoped that with Spacelab, commercial activities in space will become a reality, and that industrial companies will use and pay for Spacelab to provide important information that can be employed in making better products.

In general, a mission should involve a single scientific discipline or two closely related disciplines such as Atmospheric Physics and Earth Observations. This prevents the mission becoming too complex and hence costly. Having decided on the objectives, the best Spacelab configuration is then selected. For Astronomy this could be a pallet-only mode; for Life Sciences a module-only mode would be appropriate. The crew size, flight duration and the orbital inclination that best satisfy the mission objectives are suggested. NASA then assigns the Spacelab to a particular STS launch, deciding upon the date and which Orbiter is to be used.

The promoters of the mission then solicit experiments from researchers by sending out a formal 'Announcement of Opportunity' to all those known to be active in the field. The flight opportunity is also publicised in the technical press and by the usual 'grapevine'.

# PREPARING FOR A SPACELAB MISSION

*'Nothing in progression can rest on its original plan.'*

Edmund Burke

A microwave remote-sensing experiment was one of the many responses to the Announcement of Opportunity for the first Spacelab flight. It was chosen to fly and eventually this equipment was developed. The main reflector antenna is almost 2 metres across, and the equipment operates in the X-band at 9.65 gigahertz.

Each interested experimenter, or experimenter group led by a Principal Investigator, submits a proposal to the sponsoring body. Experiments must not only be important in their own right but also fit in with the objectives of the mission. Peer groups evaluate the scientific value of the proposals, and engineers investigate their compatibility with the Spacelab configuration to be used. Experiments which will be compatible with one another are chosen for the payload. After such screening and consideration of the political and cost factors, the payload complement is selected. Those experimenters fortunate enough to have their experiments selected prepare them in accordance with a schedule provided by a mission management team. Led by a Mission Manager, this team is itself set up by the sponsoring body. Its function is to coordinate the activities of the experimenters and provide assistance as necessary, ensuring that the hardware and software for the experiments are eventually assembled into a working payload.

## Preparation for a flight

Experiments from all over Europe are brought together and, for the first time, fitted to a pallet to form a working payload. The main equipment shown here is an X-ray spectrometer for research in Astronomy.

In order to prepare for the flight, the first task of the mission management team is to make a more detailed evaluation of the proposed payload in its relation to Spacelab. Important parameters include equipment mass and volume, resources such as power, cooling, crew attention, data rates and computer usage. This analysis ensures that the payload can be suitably accommodated and meets the basic constraints imposed by the Shuttle and Spacelab. The interactions of one experiment with another, involving factors such as heat transfer, line of sight, or electromagnetic interference, require consideration. Also, some experimenters will need a particular location within Spacelab, for example on the pallet or in racks, or use of the optical window or airlock.

A formal agreement is then made between each experimenter and the mission management to respect those constraints. While the experimenter is developing his equipment, the engineers of the management team design and develop the equipment required to help in the accommodation of the experiments, besides planning the assembly of the payload. Important agreements and specifications are formalised in documentation governing the development of the payload. These must be respected by experimenter and engineer alike.

The preparation for a flight could take anything from 2 to 5 years, depending on the complexity of the payload. Very early on, a group is set up to permit optimum sharing of the available resources by the experimenters and to solve any inter-experiment problems. This Investigators' Working Group (IWG) is made up of representatives of the experimenters who will actually be flying experiments on the mission; it ensures good cooperation between the payload management team and the investigators. The IWG normally meets about twice a year, although special meetings may be called to solve any serious problem that could affect a number of experiments. It is chaired by the Mission Scientist who is appointed by the promoting Agency to represent the experimenters to the mission management. The IWG has so far been very successful in solving problems associated with scientific performance and it is an excellent forum for the exchange of ideas.

A meeting of the Investigators' Working Group formed for the first Spacelab mission is held in Huntsville, Alabama, in September 1982, to plan how best to use the Payload Operations Control Center during the flight.

Since Spacelab is manned and flies with the Space Shuttle, all associated flight hardware must meet strict safety requirements. The mission management team therefore issues clear safety specifications for payload hardware. For example, to avoid poisoning the air and to exclude fire hazards, only certain materials can be used. The team assists the experimenter by testing his equipment as required. An experienced safety team also reviews the developed hardware from time to time and, if there is any suspicion at all that a component or procedure is not safe, changes are made.

Once the investigator has developed the hardware and software for his experiment, they are delivered to the appropriate payload management team at some central facility. There they are integrated into the payload or part payload. In the USA, this means delivery to a designated NASA Center whereas in Europe the integration centre may either be ESA-assigned or be the central facility of a certain Member State. The physical integration is usually done by an industrial team, closely supervised and assisted by the mission management team. This integration begins with joining the equipment for the experiments to the appropriate part of Spacelab such as racks and storage containers. Essential support elements are also installed such as electrical supply harnesses, cooling elements (for example, cold plates), and Remote Acquisition Units. Finally, the integration involves ascertaining that the experiments work together as a whole. This latter process includes the test and check-out of the software for each experiment which has been developed by each investigator. If this integration is performed outside the USA, the payload or part payload is subsequently delivered to the Space Shuttle launch site, Kennedy Space Center at Cape Canaveral, Florida. Sometime in late 1985 the Vandenberg Space Shuttle Launch Complex in California will become operational. This will be used for missions that require orbital inclinations greater than 57 degrees, such as the important polar and Sun-synchronous orbits.

Experiments to be performed inside the module are installed in Spacelab's racks. Here a camera–telescope, sensitive to ultra-violet radiation, is fitted in a rack for storage. When in space, it will be attached to the optical window.

The final steps in preparing Spacelab and its payload for flight take place in the Operations and Check-out Building at Kennedy Space Center. Evident in this general view are, from front to back, racks, the European experiments to be mounted on the pallet, the long module Flight Unit and, beyond the yellow arch, the Engineering Model.

When the payload arrives at the Kennedy Space Center it is delivered to the Operations and Check-out Building which, besides offices and laboratories, contains quarters for astronauts and spacecraft assembly areas. If the payload needs to be combined with another payload, the total payload is now integrated and checked out. The Spacelab configuration to be used in the flight is also prepared and all subsystems checked. Any parts of Spacelab which have already flown will have been refurbished. The next step is to integrate this payload with Spacelab. When it has been checked, using the Spacelab onboard support systems, the ensemble is loaded into a special container and towed 7 kilometres to the Orbiter Processing Facility.

As its name implies, the Orbiter Processing Facility (OPF) is where payloads are integrated into or removed from the Orbiter. Between flights the Orbiter remains in the OPF for refurbishment. This covers a complete range of tasks including the replacement of thermal tiles, subsystem repair, or even the changing of entire engine systems. On arrival at the OPF, Spacelab is placed horizontally in the cargo bay of the Orbiter. The tunnel along which the scientist–astronauts pass to enter Spacelab from the Orbiter is put into place. After check-out of the Orbiter itself and of the connections between the Orbiter and Spacelab, the Orbiter with its Spacelab cargo is taken to the Vehicle Assembly Building. There, the solid-fuel rocket boosters and the huge external fuel tanks are joined to the Orbiter. The complete assembly is then taken to the launch pad on the largest tracked vehicle in existence.

While all this hardware is being put together, the procedures to be followed during the flight are worked out. This flight operations planning is based on the needs of the experimenter, the projected flight path of the Orbiter and the capabilities of the total system. It results in the preparation of a flight 'time-line' in which specific operating periods are set aside for each experiment. Those tasks for the crew to perform must be planned and timetabled. Communications links must be established when Spacelab is in certain positions relative to the Tracking and Data Relay Satellites. The time-line is controlled by an on-board computer which has been suitably programmed.

All activities in the Orbiter Processing Facility are centred on the Orbiter. In this photograph, the three main engines of Columbia are being removed for return to the manufacturer where their performance will be improved. Brand new engines were fitted for the first Spacelab flight.

## The crew's tasks

An outstanding feature of Spacelab is that it is manned. A scientist trained in the operation of the on-board experiments is always present in the module. There are some experiments, particularly in Material Sciences and Life Sciences for which the presence of a man is mandatory. These would be carried out in the habitable module. There are others where the disturbances which might be introduced by Man make his presence undesirable. Astronomical observations, for example, require fine pointing control of the instruments. In this case experiments would be conducted on the pallet. The trained crew member, whether as experiment organiser, instrument operator or data interpreter, ensures that routine tasks are performed efficiently and that unexpected events are exploited fruitfully.

An experimenter must plan his experimental procedure as well as design his equipment. In both these areas, the knowledge of a manned presence can be advantageous. It can be assumed that the crew member will perform such menial tasks as removing equipment from stowage, setting it up, making minor repairs, changing of test specimens and disassembling equipment. Knowing this, the whole experiment procedure can be optimised. His presence is particularly attractive if 'targets of opportunity' (such as the unexpected appearance of objects in the sky or on the ground) are involved, or if readjustment or realignment on orbit are required. Indeed, for certain experiments in the Life Sciences he may act as a test subject himself.

These operations – easy for Man but difficult to achieve automatically – permit the experimenter to design relatively simple equipment. There is no need for high reliability, thereby keeping costs low. From an overall payload point of view, the crew member plays an invaluable role supervising and revising the time-line, ensuring that the experiments function correctly, and checking pallet-mounted experiments. These activities involve considerable use of the data keyboard and display systems.

Perhaps most importantly, the on-board crew member can act as a go-between, linking the experimenter on the ground with his experiment in space, by transmitting television data and by holding a conversation. Ground experimenters can thus do almost as much as if they themselves were in orbit with their experiments.

The Space Shuttle can house up to eight people in the cabin of the Orbiter. Spacelab extends the amount of living room by access to the module and its experiments. The Orbiter is flown by the Commander, who is in overall charge of the flight, and the Pilot. Both are Pilot Astronauts from the Astronaut Corps at Johnson Space Center (JSC), Houston, Texas. The duty station of the Commander and Pilot is normally the Orbiter flight deck, with full visibility of flight displays and controls. Usually, two other crew members, the Mission Specialists, also astronauts from JSC, manage the Orbiter subsystems supporting the payload. Besides this, they help the scientists operate the equipment for the experiments. It is the Mission Specialists who perform the spacewalks (Extra-Vehicular Activity, EVA) should it be necessary to visit certain places outside the Orbiter cabin or Spacelab. The rest of the crew are called Payload Specialists, and are selected for a particular Spacelab mission. They are scientists chosen for their expertise in experiments of the type carried out by

The crew of STS-9 were typical of the new blend of experience and talent needed for modern spaceflight. John Young (Commander) and Brewster Shaw (Pilot) actually flew Columbia. Astronomer Robert Parker (far right) and ionospheric physicist Owen Garriott (far left) acted as Mission Specialists. The Payload Specialists were Byron Lichtenberg (a biomedical engineer, standing left) and Ulf Merbold (a solid state physicist, standing right).

Spacelab on the particular mission. To perform these experiments is their sole responsibility; they must be able to operate the entire complement of experiments competently. The experiment crew – Mission Specialists and Payload Specialists – work shifts in the Spacelab module, usually one Mission Specialist accompanied by one Payload Specialist for a 12-hour period in the laboratory. Their off-duty hours are spent in the Orbiter. The payload crew, and sometimes the Commander and Pilot, can be subjects for biomedical experiments.

## Selection of the crew

The Orbiter Pilots and Mission Specialists are recruited by NASA and become part of the Astronaut Office at the Johnson Space Center. Each Orbiter Pilot must meet exacting medical requirements (NASA Class I) besides having had at least 1000 hours experience as a pilot-in-command of high-performance jet aircraft.

Some of the men come from the cadre of astronauts involved in the Gemini and Apollo Programs; others are newly recruited. An academic background to at least bachelor's degree standard is required in engineering, biology, physical sciences or mathematics. Although the same qualification is required of Mission Specialist candidates, much more emphasis is placed on scientific experience; indeed, a higher degree is most desirable. A pilot's licence is not required, and the medical requirements are less demanding (NASA Class II). Pilots and Mission Specialists both undergo basic training at JSC, with lectures, aircraft flights and use of flight and space simulators. After this, each astronaut is given some responsibility in some part of the Shuttle Program, and is assigned to a particular flight well in advance of the mission. From then on, he concentrates on the objectives of that flight, performing simulated missions. He also attends periodic review meetings concerned with the spacecraft or payload.

Payload Specialists are recruited for a specific mission. Candidates may be nominated by experimenters engaged in the mission or may have responded to advertisements in the press and other media. The candidates must meet basic medical criteria which are referred to as Class III Spaceflight Medical Selection Standards, but generally speaking any normal medically fit person – man or woman – is eligible. Thus, for ESA purposes, any individual who is not more than 50 years old, between 150 and 190 centimetres in height, in good health, emotionally stable and of high scientific/engineering ability can be

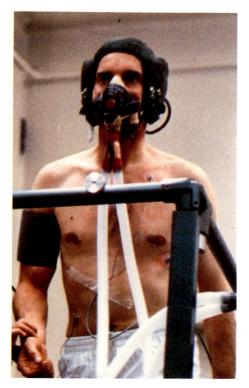

(Above) In 1977 ESA selected, from over two thousand applicants, four Payload Specialists as candidates for the first Spacelab flight. Ulf Merbold, who eventually flew on the first flight, is seen here undertaking strenuous medical tests on a 'treadmill' at the Royal Aircraft Establishment in England.

(Below) Some thirty experiments were conducted during the first Spacelab flight using the Material Sciences Double Rack. Thus, the crew had to be completely familiar with the equipment and experiments to be performed. Ulf Merbold is seen inserting a specimen into the Gradient Heating Furnace as part of his training.

considered. In fact, Spacelab was developed for normally fit scientists rather than superfit astronauts.

Those candidates who meet the medical requirements are interviewed to evaluate their scientific and technical abilities. The interviewers include members of the IWG who are representatives of the experimenters involved in the mission. The Payload Specialists chosen have proven skills and ability in the disciplines of that particular mission. They may even have unique experience of some specialised equipment to be on-board. This selection procedure is designed to give the ground-based experimenter complete confidence in the crew member who will operate his equipment in orbit. With NASA's Payload Specialists, NASA's only role is to see that the medical requirements are met; the selected Payload Specialist remains attached to his professional organisation during the preparation for the flight. European Payload Specialists become staff members of ESA. Here the final selection is made by ESA's Director General, based upon the recommendations of the Investigators.

In September 1977, fifty-three candidates were nominated for Payload Specialist positions by ESA Member States and by ESA itself out of several thousand applicants who responded to an Announcement of Opportunity issued in March 1977. All these candidates were interviewed and assessed for their suitability from a scientific, engineering and medical point of view. As a result the following four persons were chosen:

Dr Ulf Merbold of the Federal Republic of Germany
Dr Franco Malerba of Italy
Mr Claude Nicollier of Switzerland
Dr Wubbo Ockels of the Netherlands

After an initial 6 months' training period, three candidates – Ulf Merbold, Claude Nicollier and Wubbo Ockels – were retained. In view of the delays in the first Spacelab launch, NASA offered ESA's Payload Specialists the opportunity for Mission Specialist training. In 1980 Nicollier and Ockels successfully completed such training at NASA's Johnson Space Center. ESA assigned Mr Nicollier to continue with his Mission Specialist training at JSC, while Drs Merbold and Ockels became the Payload Specialists for the first Spacelab mission. These three scientist–astronauts now form the core of an ESA astronaut office which can provide experienced crew members for future ESA or ESA Member State Spacelab missions.

## Training a Spacelab crew

Once chosen, the Payload Specialists and the assigned Mission Specialists undergo training geared to the aims of the particular flight. Experience is gained in all aspects of operating Spacelab and the experiments to be carried during the flight. A very deep understanding of the workings of the subsystems and experiments is essential if the planned mission is to be successful. It also ensures that any changes of plan to accommodate unforeseen events can be made efficiently.

Payload Crew members experience two categories of training: mission-dependent and mission-independent. The longest and most important part of the training is mission-dependent; this is the part devoted to operating the experiments and to appreciating their interfaces with Spacelab. It involves having a clear insight into each experiment's methodology and an intimate knowledge of each experiment's hardware, as well as being familiar with the particular objectives and techniques of the relevant discipline. To this end, the Payload Specialists (and Mission Specialists) attend lectures, visit the home laboratories of each Principal Investigator and practise operating the experiments. Operation of the payload under realistic flight conditions is carried out in a Spacelab simulator. Such simulators have been set up in the Payload Crew Training Complex at Marshall Space Flight Center, Huntsville, Alabama, at Johnson Space Center and at the Deutsche Forschung und Versuchsanstalt für Luft- und Raumfahrt e.V. (DFVLR) establishment in Köln-Porz in the Federal Republic of Germany.

Each simulator is basically a mock-up of Spacelab, the panels within the module being copies of those in the flight unit. The crew quickly become familiar with their surroundings and the use of the keyboards and Data Display Units. Computers generate signals similar to those expected in flight using mathematical models of the experiments. The simulation director may introduce errors to the normal operations so that the crew learn how to respond to unusual situations.

Two Payload Specialists are normally trained for each Payload Specialist post on a mission. Some months before launch one of the trainees is assigned to be flight Payload Specialist and the other acts in a back-up role. If the back-up Payload Specialist is not required for the flight, he participates in the mission from the Payload Operations Control Center on the ground.

The training course ensures that the payload crew have an intimate knowledge of the operation of the experiments. It also produces an excellent team spirit among the Spacelab crew, the experimenters and the engineers supporting the mission. This team spirit was very obvious during the first Spacelab flight, when the smooth operational activities both on the ground and in space were proof of the efficiency of the training programme.

The back-up Payload Specialists followed the same training course as the flight Payload Specialists. Wubbo Ockels and Michael Lampton learn the essentials of operating the Material Sciences Double Rack.

Ulf Merbold training in the use of the Fluid Physics Module in the Spacelab one Flight Unit.

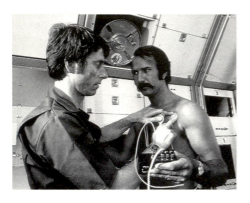

The two European Payload Specialists prepare for Spacelab's first flight. Here, Ulf Merbold fits electrodes to Wubbo Ockels to carry out medical tests.

The aim of the mission-independent training is to provide the payload crew with certain space-related skills that need to be developed for efficient operation in the space environment. Knowledge of the living and working conditions in the Orbiter and the effects of zero-$g$ are essential elements of this training which is carried out under the auspices of the Johnson Space Center.

Large cargo aircraft flying in a parabolic curve are used to achieve weightless conditions for some 30 seconds. Under such conditions, the crew learn the practical skills necessary for carrying out experiments and for living in space. They are also trained under water as SCUBA divers. Experiencing the buoyancy of the water in the one $g$ gravitational field of the Earth, they learn how to move and work in space.

Throughout the whole training, flight and post-flight periods, the Payload Specialists and their families are subject to routine medical monitoring supervised by the Crew Surgeon, a doctor assigned to safeguard the health and wellbeing of the crew. Standard medical examinations of the Payload Specialists are carried out periodically in order to guarantee that the Spacelab Payload Specialists will be completely sound in mind and body, and ready to play their very important roles in the flight.

As the time of the launch approaches, the payload crew and the experimenters take part in a grand simulation of the flight, with emphasis on the communications system. The simulation involves the crew, experiments, experimenters, control and data reception via the Tracking and Data Relay Satellite System. The exercise follows the flight time-line and all concerned – flight crew and ground personnel, including the experimenters – become familiar with the procedures to be followed during the flight. The Payload Specialists normally occupy the Spacelab simulator during this simulation while the experimenters are resident in the Payload Operations Control Center.

About 2 weeks before the launch, a Flight Readiness Review is held, attended by representatives from all the organisations involved in the mission. All flight hardware and software, plus ground equipment, and all procedures are evaluated. If all is well the decision 'to go for launch' is taken.

## Preparing for the first Spacelab flight

The Marshall Space Flight Center (MSFC) in Huntsville, Alabama, was assigned overall responsibility for the first Spacelab mission. This responsibility included overseeing the design and development of the equipment for the experiments, crew training and integration of the experiments into the payload. A Mission Manager, a Mission Scientist and engineers in a payload Project Office – all from MSFC – were appointed to coordinate all engineering and scientific aspects of the mission. NASA's Head Office in Washington, D.C., provided a First Spacelab Programme Manager and Programme Scientist. The Investigators' Working Group (IWG), whose function has already been mentioned and consisting of representatives of the experimenters contributing to the payload, judiciously guided those scientific aspects of the developing payload through its difficult gestation period and successfully resolved conflicting demands made upon Spacelab resources by individual experimenters.

Another area where the advice of the IWG was essential concerned the decision to go for launch with only a single satellite of the Tracking and Data Relay Satellite (TDRS) System in orbit. It was originally planned that two such satellites would be available for the mission, and the Principal Investigators planned their experiments accordingly. However, due to the STS-6 IUS (Inertial Upper Stage) launch problem, only one such satellite was in orbit at the time of the mission. At the IWG meeting in March 1983, a thorough analysis of the situation regarding data transmission and command was made by members of the IWG, and by NASA and ESA engineers. It was agreed that a '30 September 1983 mission with one fully operational TDRS was scientifically viable'. Although the originally hoped-for scientific return could not be fully realised, for most experimenters one TDRS was sufficient under normal conditions; of course the risk of data loss during unusual conditions was considerably increased.

Although the overall responsibility for the mission rested with NASA, ESA's contribution was managed by a First Spacelab Payload (FSLP) Programme Manager, and the science was supervised by a FSLP Project Scientist. In 1977, ESA set up a small management team called SPICE (Spacelab Payload Integration and Coordination in Europe) to plan and assemble the experiments destined for the first Spacelab flight into a complete payload complement. This team was based at the DFVLR establishment in Köln-Porz in the Federal Republic of Germany; it received technical support from DFVLR and from ESA's technical and operational centres at Noordwijk (the Netherlands) and Darmstadt (the Federal Republic of Germany). Also, the French Space Agency, Centre National d'Etudes Spatiales (CNES), provided valuable assistance in integrating the experiments originating from France. The tasks of the SPICE team are summarised as follows:

ESA's scientist–astronaut Claude Nicollier trains for a Life Sciences experiment that was performed on the first Spacelab flight. Here, he experiences a short period of zero-*g* aboard a special KC-135 aircraft.

  (i) Studying the accommodation of the selected experiments in Spacelab and coordinating the experiment activities
  (ii) Issuing payload specifications and ensuring that these are met
 (iii) Providing technical assistance to experimenters during all phases of the mission
 (iv) Procuring the required support hardware, including the necessary test equipment
  (v) Developing the ground check-out procedures and on-board software
 (vi) Planning, defining and managing the physical integration of the European portion of the first Spacelab payload
 (vii) Defining the flight operations associated with operating the FSLP in space
(viii) Supervising the training of the Payload and Mission Specialists in Europe
 (ix) Providing support to NASA before, during and after the first Spacelab flight.

The major contractor for the physical integration and testing of the FSLP experiments was the MBB–ERNO company at Bremen. The payload was assembled in ERNO's Integration Hall previously used for the integration of Spacelab itself. After the integration activities were completed in Europe, in May 1982, the ESA portion of the payload was flown to the Kennedy Space Center in Florida. There the ESA and NASA portions were brought together to form the first Spacelab payload.

| MAJOR MILESTONES | 1976 | 1977 | 1978 | 1979 | 1980 | 1981 | 1982 | 1983 |
|---|---|---|---|---|---|---|---|---|
| *(milestones)* | ↑ Announcement of opportunity | ↑ SPICE formed · ↑ Experiments selected · ↑ Payload Specialists selected | | | ↑ Start payload integration | | ↑ FSLP delivered | ↑ ESA–NASA payload integrated · ↑ Launch |
| Experiment definition, development and acceptance | | ▓▓▓▓▓▓▓▓▓▓▓▓▓▓▓▓▓▓▓▓▓▓ | | | | | | |
| Accommodation analyses | | ▓▓▓▓▓▓▓▓▓▓▓▓▓▓▓▓ | | | | | | |
| Integration (Europe) | | | | | ▓▓▓▓▓▓▓ | | | |
| Integration (KSC) | | | | | | | ▓▓▓▓▓▓▓ | |
| Crew training | | ▓▓▓▓▓▓▓▓▓▓▓▓▓▓▓▓▓▓▓▓▓▓▓▓▓▓▓▓▓▓▓▓ | | | | | | |

The European contribution to the first Spacelab payload was developed during the late 1970s. Under the management of SPICE, the individual experiments were integrated into a payload and later joined with the NASA payload in readiness for the flight in November 1983.

In parallel, the flight elements of Spacelab's Flight Unit were being assembled and checked out in preparation for the flight. During early 1983, experiments in their appropriate racks were integrated into the module, and the pallet experiments were mounted. In the latter case the pre-integration had taken place using support structures. These structures were used to ensure better viewing for the experiments by raising the equipment to the top of the pallet. This permitted the installation of cold plates, RAU's, etc. under the structures and also allowed integration of the ESA and NASA portions to proceed independently.

The physical integration of ESA's First Spacelab Payload experiments was carried out by MBB–ERNO. In the Integration Hall at Bremen, Federal Republic of Germany, final adjustments are made to the pallet-mounted experiments.

ESA's experiments to be mounted on the pallet
are inspected on arrival at Kennedy Space
Center.

Those experiments requiring direct exposure to
space are mounted on the pallet. The
European experiments are on the left and the
US experiments on the right. The red tags
indicate instruments that require further
attention and are removed before flight.

When all the flight elements of the first Spacelab mission had
been assembled – the module with its support equipment, the pallet,
the payload, the electrical power harnesses and structural interface
fittings – a fit check was made using the Cargo Integration Test Equipment (CITE). The CITE, an exact replica of Columbia's cargo bay,
ensures that Spacelab, with its tunnel and payload, fits into the Orbiter
cargo bay and makes the necessary connections correctly. This check
was completed on 1 July 1983.

The racks used to house many of the first Spacelab experiments are swung into position for integration into the module.

The core module is prepared to receive Spacelab's subsystems and rack-mounted experiments.

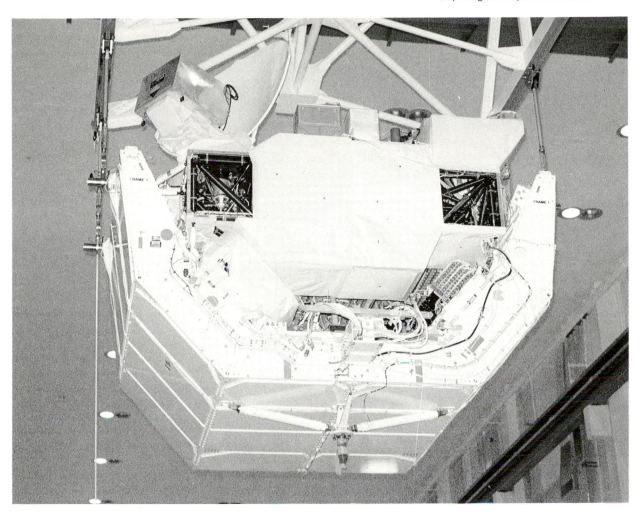

The pallet and its instruments are moved to join the module.

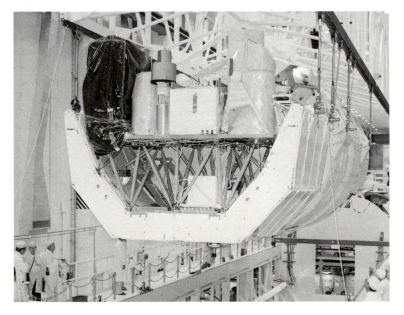

The Spacelab module and its attached pallet are lowered into a full-scale model of the Orbiter's cargo bay (CITE) for the critical cargo integration tests.

The Spacelab module and pallet rest in the
CITE stand.

Spacelab, packed with its experiments, is
hoisted from the CITE fixture to the
environmentally controlled canister that will
take it to the Orbiter Processing Facility.

Spacelab is lifted out of the canister into Columbia's cargo bay in the Orbiter Processing Facility in August 1983.

The complete Spacelab is installed in Columbia's cargo bay. The access tunnel is particularly prominent in this photograph. All exposed surfaces are covered with white blankets for thermal insulation.

Tests were carried out to make sure that all equipment was working perfectly. Selected parts of the flight time-line were operated and supported by actual Spacelab subsystems, in contrast to earlier tests which had merely simulated those subsystems. Those instruments scheduled for operation during those parts of the time-line were switched on, and the compatibility of each with the other experiments to be operated simultaneously was verified.

A big day for Spacelab and its payload occurred on 25 July 1983 when the total cargo was officially declared ready for flight. In the Cargo Readiness Review held on that date, NASA and ESA examined the appropriate documentation and the results of all the tests. The precious cargo was taken on 16 August 1983 to meet Columbia in the Orbiter Processing Facility, where the final stages of integration were completed. Spacelab was lowered into the cargo bay of Columbia, and the access tunnel installed.

From 7 to 9 September 1983, an important test of the total communications system took place. This so-called 'end-to-end' test involved all elements of the data flow, that is from Spacelab to Orbiter, TDRSS, DOMSAT (a US domestic communications satellite system used in this case for transferring data from White Sands to the POCC), Mission Control Center and POCC. The tests confirmed that all types of data – real-time, digital, playback of the on-board recorder (HDRR), analogue and video – could be transmitted from Spacelab to the experimenters and engineers on the ground.

While the flight hardware was being integrated, crew and experimenters alike were preparing themselves for the operation of the experiments. A time-line was prepared that detailed every step to be followed during the flight. Based upon the requirements of each experimenter, this took into account such factors as Orbiter orientation and position relative to the Earth's surface, visibility of the TDRS

Spacelab first encounters Columbia in the Orbiter Processing Facility and the tunnel linking the two is fitted.

from the Orbiter, data transmission and recording, the position of the ground and space targets to be observed, and off-duty activities of the crew such as sleeping and eating. The time-line programmed into Spacelab's Command and Data Management Subsystem is geared to 'mission elapsed time', that is the time measured from the time of lift-off. It would represent a 'road-map' for the flight activities; information contained in it could be displayed to the crew at any time.

The operations to be conducted in accordance with this time-line were practised by the payload crew and experimenters in August and September 1983. The payload crew used the Spacelab simulator at Huntsville, the experimenters and the flight operations support teams being in the POCC at Houston with their Ground Support Equipment. For the longer (dress rehearsal) simulations, Columbia's Pilots in the Orbiter simulator at Houston joined in the activities. The overall control of the mission was conducted from the Mission Operations Control Room (MOCR), also at Houston. All of the computers, displays and data links to be used during the actual mission were employed.

The planned launch date of 30 September 1983 was changed to 28 October 1983 due to uncertainties in the operation of the single Tracking and Data Relay Satellite available. The arrival of a hurricane in the Houston area caused further complications! Then a problem was encountered with the solid-fuel rocket boosters (SRBs) and the launch was further delayed. Examination of the SRBs used for the STS-8 mission showed excessive erosion of the lining to the booster's nozzle. The Shuttle–Spacelab combination already on the launch pad and being prepared for launch was rolled back into the VAB on 17 October 1983. The problem was traced to a change in the procedure for manufacturing the lining. It was cured by replacing the suspect right-hand solid-fuel rocket with one made before the change took place.

Meanwhile, the Orbiter and Spacelab were removed from the Space Shuttle stack and returned to the OPF for safe keeping. Items such as fuel cell parts, batteries and films – with limited lives – were replaced. The liquid nitrogen supply for one experiment was topped up. When the go-ahead was given for launch the Orbiter–Spacelab combination was towed back to the VAB and again mated with the SRBs and large external tank containing propellant. The complete Space Shuttle was rolled out to the launch pad again on 7 November to be prepared for its launch on 28 November 1983. The November launch date was agreed jointly by ESA and NASA only after considerable deliberation.

The launch date and time depend on a number of factors that influence the scientific results that can be obtained. The time of the year influences Earth observations since good photography of the Earth's surface is only obtained if the Sun angle is advantageous and the target is not covered by cloud, fog or snow. A launch near New Moon is essential so that the level of background light is low enough to observe sources of low intensity in Astronomy and Atmospheric Physics experiments.

Two conflicting requirements had to be taken into account determining the time of day for launching Spacelab. An early daytime launch in Florida would allow an emergency landing in daylight at Zaragoza, Spain, immediately after launch if necessary. On the other hand, a later launch from the Kennedy Space Center would give more time in the Earth's shadow. The agreed compromise launch time was

NASA's Minilab that houses equipment for five Life Sciences experiments is transported by crane to join the group of racks destined for the Spacelab module.

11.00 Local Time in Florida (16.00 GMT), with a launch window of only 14 minutes.

The final decision to commit to launch on 28 November 1983 was taken by a NASA/ESA management team at a Flight Readiness Review on 18 November. Even so, it was recognised that seven of the experiments on-board would suffer a reduced scientific return due to this late-in-the-year launch. For these, a free reflight on later missions was guaranteed. The countdown to launch started on 25 November 1983.

Thus with the major disadvantages of a late launch and the fact that only one TDRS was available for communications the first Spacelab flight was made under far from ideal conditions. The fact that significant results have been returned from the mission is a tribute to the experimenters and mission support teams on both sides of the Atlantic.

All is ready for the launch of Spacelab. The Space Shuttle Columbia, with its Spacelab cargo, receives its final check at dawn on 28 November 1983.

## Highlights

Spacelab flew for the first time on 28 November 1983. It was launched in the cargo bay of the Orbiter Columbia, exactly on schedule, at 11.00 Local Time (16.00 Greenwich Mean Time, GMT), from the Kennedy Space Center in Florida. This was the ninth Space Shuttle flight, designated STS-9. Columbia, with Spacelab contained in its cargo bay throughout the flight, completed 166 orbits of the Earth; the historic flight lasted 10 days, 7 hours and 47 minutes. Columbia landed at the Edwards Air Force Base in California on 8 December 1983 at 23.47 GMT. At almost 100 000 kilograms, this was the heaviest vehicle returned to Earth from space so far. The mission was a great success.

The main objectives of the flight were to show that Spacelab works, and to demonstrate its use as a space laboratory over a wide variety of scientific and technical disciplines. Other objectives were to investigate the local environment caused by Spacelab itself and to prove that all the connections between Spacelab and the Orbiter performed correctly. The Spacelab configuration used was the long module plus one pallet which, with its payload, had a mass of about 15 000 kilograms.

# THE FIRST SPACELAB FLIGHT

*'Hail Columbia! happy land!*
*Hail, ye heroes! heaven-born band!'*
Joseph Hopkinson

Columbia lifts off at 16.00 GMT on 28 November 1983.

Ulf Merbold waves farewell as members of the STS-9 crew enter the van that will take them to the launch pad.

The scientific objectives of the first Spacelab mission were chosen jointly by ESA and NASA to demonstrate Spacelab's use in a variety of disciplines. Briefly stated, they are as follows:

(i) Studies of ultra-violet and X-radiation sources in the Universe (Astronomy)

(ii) Precise measurements of the Sun's energy output (Solar Physics)

(iii) Studies of the Earth's plasma envelope and of the environment around the Space Shuttle (Space Plasma Physics)

(iv) Study of the Earth's environment, through investigations of atmospheric composition and motions, and the emission and absorption of radiation of different wavelengths (Atmospheric Physics)

(v) Demonstration of advanced remote-sensing systems (Earth Observations)

(vi) Observation of microgravity effects on the solid, liquid and vapour phases of different substances; various experiments in fluid physics, crystal growth and metallurgy (Material Sciences)

(vii) Evaluation of tribology (that is, friction) under microgravity conditions (Technology)

(viii) Study of the effects of the space environment on human physiology and on the organisation of biological systems (Life Sciences).

The orbit was circular, at 240 kilometres above the Earth, and the inclination of the orbit was 57 degrees so that Spacelab overflew the USA and most of Europe. The 57 degrees North latitude line represents the northern limit reached by the Orbiter–Spacelab configuration. Similarly, the southern limit of the ground track is represented by the 57 degrees South latitude line, so that Spacelab flew over the whole of Australia, New Zealand and South America. Some two hundred attitude manoeuvres were made by firing Columbia's attitude control motors in order to put the Orbiter–Spacelab duet through its paces and to carry out the exacting flight programme. For the first Spacelab mission there were six crew members aboard Columbia.

## The crew

John Young was the Commander, Brewster Shaw the Pilot, Drs Owen Garriott and Robert Parker the Mission Specialists, and Drs Ulf Merbold and Byron Lichtenberg the Payload Specialists for the first Spacelab mission. Drs Merbold and Lichtenberg, the first of the new breed of astronaut–scientists, were specially picked for their scientific abilities and their skills in operating the experiments to be conducted on this flight. Dr Merbold was the first non-American to fly on the Shuttle. The first ESA member to fly in space, he has inaugurated a new era of European space exploration – European manned spaceflight.

At the time of the first Spacelab flight and aged 54, John Young had over 800 hours of space experience. He is the first – and only – man to have made six trips into space; he flew on two Gemini missions (flights 3 and 10), flew on two Apollo missions (10 and 16) to the Moon, and commanded the first Shuttle flight STS-1. He was the ninth man on the Moon and a driver of the Lunar Roving Vehicle. Recruited by NASA in September 1962, he is now Chief of the Astronaut Office at the Johnson Space Center. His wide experience proved to be an invaluable contribution to the success of the STS-9 mission. John Young's pilot on this flight was Brewster Shaw, making his first spaceflight. At the age of 38 he was the youngest member of the Orbiter crew. He holds a BS degree in engineering, and has been a test pilot and instructor in the United States Air Force.

The two Spacelab Mission Specialists were Owen Garriott (then aged 53) and Robert Parker (47). Both have strong connections with the earlier Skylab Program; launched on 14 May 1973, Skylab was the USA's first Space Station. It carried three different three-man crews who in all completed about 3900 orbits of the Earth at 430 kilometres altitude and 50 degrees inclination. Dr Garriott, a holder of BS, MS and PhD degrees, was an Associate Professor of Electrical Engineering at Stanford University during the early 1960s. Selected

(Above) Owen Garriott enters Spacelab through the tunnel from Columbia to start Spacelab one's research programme.

(Below) When a satellite orbits the Earth at about 250 kilometres altitude in an orbit inclined at 57 degrees (as in the Spacelab one flight), its track on the Earth's surface is as shown here. Successive tracks cover different parts of the Earth since, during one satellite orbit, the Earth rotates through an angle of 22.5 degrees.

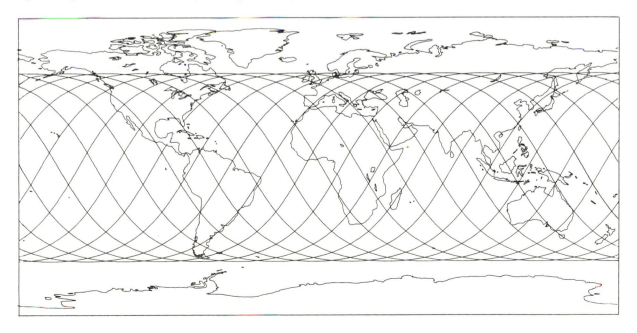

This striking picture of the rising Sun and the
glow of the Earth's atmosphere was taken from
the window of the aft flight deck of Columbia
whose surfaces are bathed in the first light.

by NASA as an astronaut in 1965, he served as science-pilot on the second manned Skylab mission, logging 59 days in space, of which 13
hours and 43 minutes were spent on Extra-Vehicular Activity (EVA)
outside the spacecraft. Dr Parker, Program Scientist for the Skylab
Program, has also had considerable experience in support of the
Apollo Program. However, STS-9 was his first flight in space. An
astronomer by profession, holding BA and PhD degrees, he was an
Associate Professor of Astronomy at the University of Wisconsin before
joining NASA in 1967.

Dr Ulf Merbold, ESA's Payload Specialist for the first Spacelab
mission, was born in June 1941 at Greiz in Germany. He is a solid-state
physicist from the Max Planck Institute for Metals Research, where his
main fields of research were crystal lattice defects and low-tempera-
ture physics. He holds a Diploma in Physics and a Doctorate in Science
from Stuttgart University. He joined ESA in December 1977. The US
experimenters were represented by Dr Byron Lichtenberg, a 35-year-
old biomedical engineer from the Massachusetts Institute of Technology. Still a member of the research staff at MIT, he was selected for
Payload Specialist training in June 1978. He received his BS degree
from Brown University and his MS and PhD degrees from MIT. He is
an ex-USAF pilot who was awarded two Distinguished Flying Crosses
during his tour of duty in Vietnam.

The two flight Payload Specialists were nominated for the mission in September 1983. At the same time two back-up Payload
Specialists, Dr Wubbo Ockels for ESA and Dr Michael Lampton for

NASA, were assigned. Dr Ockels, then 37 years old, is a physicist from the Netherlands; he holds a PhD degree in nuclear physics from the University of Groningen. He has successfully completed the NASA Mission Specialist training course at NASA's Johnson Space Center in Houston. Dr Lampton, aged 42, is a research physicist working in astronomy and space physics at the Space Sciences Laboratory of the University of California, Berkeley, where be obtained his PhD degree.

The back-up Payload Specialists underwent the same training as those who flew in space and were equally well qualified in the operation of all the experiments. In addition to being back-ups for the flight, Drs Ockels and Lampton were deeply involved from the ground in payload operations throughout the mission. They acted as the bridge linking the experimenters on the ground and the on-board crew. Their intimate knowledge both of the experiments and of Spacelab itself was invaluable. They also acted as test subjects (together with the flight Payload Specialists) in some Life Sciences experiments to provide ground control data.

During the first Spacelab flight, the crew worked in two teams, in 12-hour shifts. The Red Team consisted of Young, Parker and Merbold, while the Blue Team contained Shaw, Garriott and Lichtenberg. This Pilot–Mission Specialist–Payload Specialist combination ensured that the Orbiter, Spacelab and its experiments were operated in unison at all times.

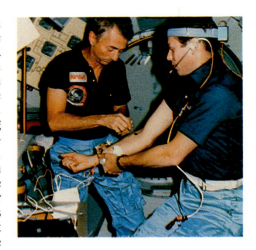

(Above) The scientist–astronauts took part in medical experiments to determine the effect of near-weightlessness on the distribution of body fluids, the circulation of red blood cells and the body's ability to resist disease. Here, Owen Garriott is measuring Byron Lichtenberg's venous pressure and taking a blood sample.

The crew's first glimpse of Earth from orbit at the start of the first Spacelab flight.

A space portrait of the Spacelab one crew showing the Red Team, John Young, Robert Parker and Ulf Merbold (at the top) and the Blue Team, Brewster Shaw, Owen Garriott and Byron Lichtenberg (at the bottom).

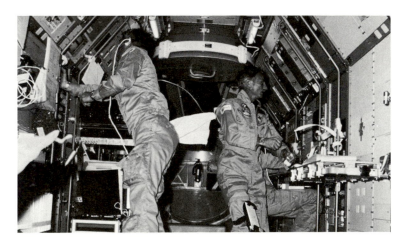

The Mission Specialists, Owen Garriott and Robert Parker, work in Spacelab's module.

## Validation of Spacelab's performance

During the development of Spacelab no tests of the system were conducted in space before the first mission. This money-saving approach meant that all subsystem and system tests had to be carried out under laboratory conditions on the ground. Although the tests were exhaustive and a high degree of confidence in the equipment existed, it was essential to evaluate Spacelab's performance under orbital flight conditions. Hence, the primary purpose of the first Spacelab mission was to prove the efficient operation of its thousands of structural, mechanical and electronic components. Also, Spacelab had to be shown to be compatible with the Orbiter and to be a laboratory in which a broad range of experiments could be conducted successfully. Since Spacelab is a base for making scientific measurements, it was also necessary to know if its own environment could affect the observations made by the instruments that it carried.

Throughout the first Spacelab flight, measurements of temperatures at particular places, of pressures in the module or in tanks holding fluids, and of electrical voltages at selected points of the avionics system (Spacelab's entire electrical and electronic system) were taken on a routine basis. Such standard operational measurements were required by subsystem designers to verify their designs. In addition, a Verification Flight Test (VFT) programme was conducted by NASA's Marshall Space Flight Center to evaluate the overall system performance in essential areas. These concerned the structure, Spacelab's facilities for the crew, environmental control, electrical power distribution, command and data management, properties of Spacelab surface materials, and the environment outside Spacelab. For this programme, the standard Spacelab measurements were used, as were measurements from Verification Flight Instrumentation (VFI) specially installed as part of the Verification Flight Test programme. Observations were made during ascent, on-orbit and during re-entry.

## The joint European–US scientific payload

ESA and NASA devised a joint payload for this first Spacelab mission. Each Agency was allocated about half of the available resources. The scientific payload had a mass of 2785 kilograms; 2 kilowatts of power were allocated for its operation. The energy consumed by the experiments was planned to be about 170 kilowatt hours. These resources were less than the normal Spacelab allowance due to the experimental nature of the mission and because other instrumentation (such as the 856-kilogram Verification Flight Instrumentation, VFI) was also being carried on-board.

ESA and NASA issued separate Announcements of Opportunity seeking experiments to satisfy the mission objectives. Each Agency then selected its quota of experiments according to their scientific merit and suitability for flight on the Shuttle. Out of some four hundred proposals, seventy experiments were chosen from eleven European countries, the USA and Japan. The experiments chosen

Byron Lichtenberg, the US Payload Specialist, checks the Material Sciences Double Rack before performing one of the thirty experiments carried out using this facility.

Ulf Merbold, ESA's Payload Specialist, inserts a sample contained within a cartridge, to be treated in the Gradient Heating Furnace of the Material Sciences Double Rack. His posture shows that, even in zero-*g*, the body tends to take a sitting position.

were from the disciplines of Astronomy, Solar Physics, Space Plasma Physics, Earth Observations, Material Sciences, Technology, and Life Sciences. The main emphasis of the experiments in the NASA part of the payload was on phenomena occurring in the atmosphere, whereas Material Sciences formed the main theme for the ESA complement.

The European experimenters were themselves responsible for the planning and funding of their experiments, and also for the development of their equipment. (In contrast, US experimenters developed their equipment under contract from NASA.) Due to the centralised nature of the data handling, experiment-related software (Experiment Computer Application Software, ECAS) was developed jointly by teams from ESA and DFVLR. Compatibility with the Experiment Computer Operating System (ECOS) developed by NASA was thus assured.

The Spacelab one time-line illustrates the major happenings of the 10-day flight. Of particular interest is the three-way link-up between the crew on-board Spacelab with President Reagan (of the USA) in Washington and Chancellor Kohl (of the Federal Republic of Germany) who was in Athens at that time.

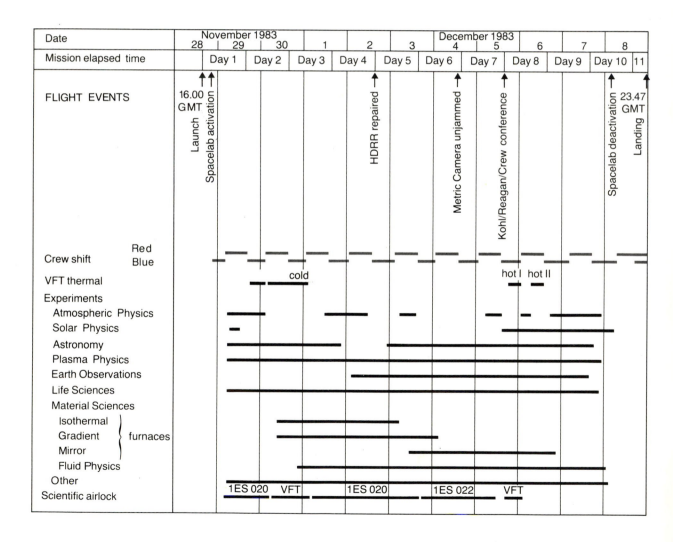

(a)

(b)

(c)

(d)

These photographs of Earth were taken from Columbia during the Spacelab one flight using a hand-held Hasselblad camera. They show (a) Cape Fear and Wilmington, North Carolina, USA; (b) Kamchatka Peninsula, USSR; (c) Tenerife and Gomera, Canary Islands, Spain; and (d) Ras Al Hadd, Muscat, Oman.

(Above) From the Payload Operations Control Center in the Johnson Space Center, Houston, Texas, investigators with experiments on Spacelab one could discuss the operation of their experiments in real time with the astronauts on-board.

(Below) The STS-9 flight was controlled overall from the Mission Operations Control Room adjacent to the POCC. There, all aspects of the Columbia–Spacelab flight were monitored and the Flight Director could provide instructions to Commander John Young based on all the factors influencing the mission. In the foreground of this picture, the ESA Crew Surgeon can be seen at his console.

# Operations during the first Spacelab flight

The flight progressed according to plan; all the major objectives were accomplished. After a flawless launch, Spacelab was activated after 2.5 hours and the Payload Crew entered the laboratory one hour later. Operation of some experiments started about 4.5 hours into the flight. During the first day, the crew worked a little more slowly than anticipated and there were a few delays in events scheduled on the time-line. Even so, most planned activities were completed. Also, there were a few minor problems with those experiments controlled by the computer; some observation opportunities were missed due to TDRSS problems. After these initial difficulties had been overcome, everything proceeded smoothly. The activity time-line was optimised from time to time, based on the scientists' interpretations of their results. Since the rate at which the on-board consumables were used was quite low, an additional day in space was permitted. This extra day was used to compensate for some early losses in data taking and to extend the scope of certain experiments.

The STS-9 flight was controlled from the Mission Operations Control Room (MOCR) in Houston, Texas. There the Flight Director and his supporting personnel experienced in such fields as navigation, payload performance and crew health, maintained round-the-clock contact with the Orbiter. The hub of the experimenters' involvement was the Payload Operations Control Center (POCC). Being in the same building as the MOCR, direct communications were possible between the two groups. During the whole flight, the Principal Investigators, with their experiment teams and Ground Support Equipment, were at hand in the POCC as the time-line unfolded.

The back-up Payload Specialists worked two 12-hour shifts in the POCC and acted as voice links with the crew aboard Spacelab. Audio and television links (via the TDRS) enabled the scientists to follow the progress of their experiments and to be intimately involved as they progressed. Data received could be displayed and/or recorded and some on-the-spot data processing was done. The Spacelab crew performed their part of this combined operation extremely well. They acted as a true 'alter ego' of the experimenter on the ground and proved themselves to be absolutely indispensable. Since nearly all the Material Sciences experiments involved European experimenters, the data in this discipline were transmitted in real time to DFVLR at Oberpfaffenhofen in the Federal Republic of Germany. Some experimenters thus successfully monitored the progress of their experiments without travelling to Houston.

In addition, NASA and ESA engineers were present to provide the necessary management and analysis support in the POCC. The performance of Spacelab itself was monitored and evaluated by a NASA/ESA/contractor team situated at Huntsville, Alabama, and termed the Huntsville Operations Support Center (HOSC). There, real-time data from Spacelab during all phases of its flight and for all the relevant payload operations were presented on consoles that were manned throughout the mission. In this way the HOSC team could closely watch temperature, pressure and electrical measurements, and thus perform part of the Verification Flight Test (VFT) programme. The HOSC was linked directly with the MOCR and POCC so that personnel were able to observe in-flight activities by means of closed-circuit television. Using the Spacelab mock-up, engineers were able to solve problems as they arose.

The landing of Columbia was delayed by about 8 hours when problems were encountered with two of the five on-board computers and with one Inertial Measurement Unit which provides navigational information. Columbia with Spacelab, landed at Edwards Air Force Base, which adjoins the Dryden Flight Research Center in California. Immediate post-flight operations involved ensuring that the Orbiter was safe and removing delicate items of experimental equipment – exposed films, some biological samples and instrumentation urgently needed for later missions. After the successful re-entry and landing, it was discovered that there had been a small fire on-board, associated with an auxiliary power unit used on the approach to landing. However, these minor problems did not detract from the tremendous accomplishment of STS-9. The Orbiter was then returned to the Kennedy Space Center by means of the special Boeing 747 aircraft. There, Spacelab was removed from Columbia; the payload was disassembled by mid-February 1984, and all equipment returned to its owners.

Immediately after landing, the Payload and Mission Specialists were subjected to medical tests at Dryden as a continuation of the in-flight Life Sciences experiments. These lasted 7 days and included zero-$g$ flights in aircraft and runs on a mechanical sled during which they experienced controlled accelerations. Thus results gained before and during the flight could be compared with those made afterwards.

Tired – but happy – the STS-9 crew, led by Commander John Young, are welcomed back to terra firma.

(Above) The crew badge for the STS-9 flight depicts the Orbiter Columbia carrying Spacelab, for the performance of Spacelab one. The badge was worn by all crew members during the flight.

(Below) Columbia, on the runway at the Edwards Air Force Base, California, after its return from space, is being prepared for its piggy-back trip back to its launch site at

Analysis of the data returned from the flight will take many months. All the television and voice information generated during the mission was recorded; in addition some experimental observations were recorded on-board. All in all, some 1500 magnetic tapes were used. The tapes, edited and checked at NASA Goddard Space Flight Center and DFVLR in Europe, contained some additional information on the Orbiter's position and Spacelab 'house-keeping' data, such as temperatures measured at certain places. These tapes have been delivered to designers for evaluating Spacelab's performance and to investigators for analysing their experimental results.

The first Spacelab mission cannot be considered to have been completed until all the results obtained have been published. As far as the design teams are concerned, this means that the data must be analysed and the 'off-nominal' conditions evaluated. The final act is to improve a design or a procedure so that later missions will be carried out under even better operational conditions. For the scientists, the publication of their results is their final goal. However, these results are not an end in themselves but provide the investigators with some information on pieces of the puzzles that fascinate them. Spacelab is available to help further as and when required.

Kennedy Space Center, Florida, atop a Boeing 747. After refurbishment and change of payload, it will be ready for another flight into space.

## Proof of Spacelab's performance and environment

The primary objective of the first Spacelab mission, that is, the proof of Spacelab's performance in orbit, was achieved in all respects. To do this a multitude of measurements at key points had to be made and their significance assessed.

The structure received considerable attention because structural integrity is closely linked with the basic safety of Spacelab and the number of times that it may be re-used, that is, its operational lifetime. Spacelab's pressurised module was found to maintain its nitrogen and oxygen atmosphere perfectly – not a single leak was noted. The scientific airlock has moving mechanical parts; since the airlock is the connection between the pressurised laboratory and the vacuum of space, it is a very critical item. Its performance was outstanding during six completed cycles of operation.

The Spacelab window in the ceiling of the core segment, to which the Metric Camera was attached, performed well, although some particulate matter appeared on its outer surface. Also, the good optical quality of the triple-layer glass of the two viewports was maintained throughout the flight.

Some fifty strain gauges were used to measure the deformation of Spacelab itself and of its attachments to the Orbiter. Both the axial and the rotational responses of the structure to low-frequency loads imposed throughout the flight were monitored by some thirty accelerometers placed at suitable positions on the tunnel, module and pallet. Measurements of the variations in acceleration with time provided data on load cycles and metal fatigue. The dynamic response of structures of relatively large mass, such as rack-to-floor and pallet-to-Orbiter connections were included in these accelerometer measurements. Module, pallet and tunnel responses to shock and random vibration were monitored over a wide range of frequencies. Acoustic loads imposed during launch, and caused by rocket exhaust noise, were deduced from microphones placed inside and outside the module, and on the tunnel and pallet. The measured strains and accelerations were within the ranges predicted before the flight.

# PRELIMINARY RESULTS OF THE FIRST MISSION

'One must learn by doing the thing; though you think you know it, you have no certainty until you try.'

Sophocles

The airlock, used to place into space instruments that are stored in the module, worked perfectly during the first Spacelab flight.

The value of the acceleration within Spacelab is not exactly zero due to the motions of people and machinery. The residual acceleration levels were measured, on-orbit only, by twelve sensitive accelerometers. Responding to frequencies up to 20 hertz, these were strategically placed in the module and on the pallet. It was found that the actual residual acceleration was generally between $10^{-4}g$ and $10^{-5}g$, with occasional peak values of $10^{-3}g$.

The radiation level within Spacelab, due to X-radiation, gamma-radiation, neutrons and cosmic rays, can affect not only the health of the crew but also sensitive films, biological samples, integrated circuits and the interpretation of experimental results. It was important to show that the radiation inside Spacelab was relatively low, and predictable. Measurements were made using standard dosimeters, nuclear emulsions and silver chloride crystals. Proportional counters measured the variations of charged particle fluxes with time. The observations were transmitted to the ground by the data link and recorded for later detailed analysis.

No loose items were found floating around Spacelab when the crew entered it 3.5 hours after lift-off. The positioning of the handrails and the containers was found to be satisfactory, but the elastic cords used as stowage restraints were too strong. For future Spacelab flights, less strong elastic straps and more double-sided Velcro material are suggested for keeping the many bits and pieces used in the laboratory conveniently to hand. These conclusions were reached

Measurements were made throughout the module, pallet and tunnel as part of the Verification Flight Test programme.

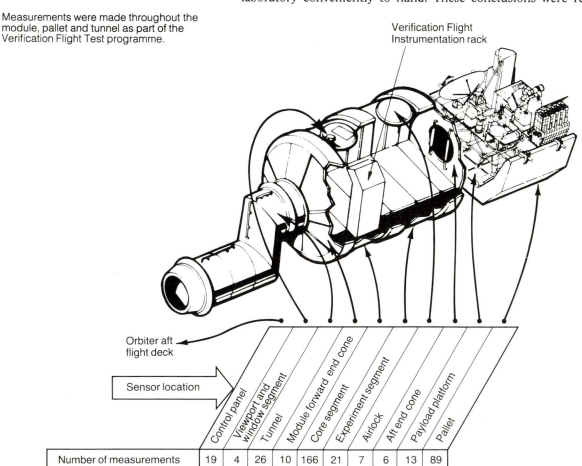

Verification Flight Instrumentation rack

Orbiter aft flight deck

Sensor location

| | Control panel | Viewport and window segment | Tunnel | Module forward end cone | Core segment | Experiment segment | Airlock | Aft end cone | Payload platform | Pallet |
|---|---|---|---|---|---|---|---|---|---|---|
| Number of measurements | 19 | 4 | 26 | 10 | 166 | 21 | 7 | 6 | 13 | 89 |

from crew transcripts and comments, and from video recordings showing the crew performing such activities as removing equipment from storage containers.

A portable sound level recorder measured the noise levels at six strategic points in the module. These measurements showed that the laboratory was a quiet place in which to work. Its acoustic environment was most acceptable to the crew for the many types of activities required for future Spacelab flights. Operation of the fans and pumps aboard Spacelab was inaudible. The High Data Rate Recorder (HDRR) produced some noise during fast wind operations, but this was only a problem in the vicinity of the recorder itself. Occasionally, the crew heard 'bangs' during the flight. These have been attributed to temperature changes of the access tunnel; some central heating pipes make noises in much the same way. Also the astronauts reported that the intercom was not quite loud enough.

Air samples were taken from inside the module periodically during the flight; they are being analysed for their toxicity. No unpleasant odours were reported in the module. Only when the canister of lithium hydroxide, which efficiently removed carbon dioxide from the air, was changed and when the valve regulating the flow of nitrogen into the module was manually controlled, were warning alarms raised. This was expected and proved the effectiveness of the alarm system. Thus the safety of Spacelab was extremely well demonstrated by this first mission.

For further flights involving experiments on the behaviour of liquids in space, a sponge will be required to collect spilt liquid. In the first Spacelab flight, such experiments caused the only three boxes of tissues available to be completely used! The tidying up took an hour!

The Environmental Control Subsystem (ECS) was checked in two ways. One test verified the design of the passive thermal control elements through temperature measurements taken at various places throughout the module and on the pallet. Many Spacelab attitudes – directly facing the Sun, looking at deep space, or pointing towards the Earth – were investigated so that hot and cold conditions could be experienced and the corresponding temperatures measured. Even so, extreme design conditions could not be reached during this flight. However, the test results can be used to prove the mathematical model, which is implemented as a computer program, and used to predict temperatures at all points under all conditions. The other test was designed to evaluate the performance of those elements of the ECS which control the crew's life support systems and equipment. This involved the measuring of pressure, temperature and humidity within the module. The relative humidity was kept below 40%, the air temperature was about 20 degrees Celsius and no water condensation was observed. Thus, a most comfortable environment is provided for working in the space laboratory.

Over 300 sensors measuring temperatures, pressures, gas flow rates, smoke presence and control valve operation were involved in these ECS tests. Sensors were placed in the tunnel, on the end cones, at various positions within and outside the module, and on the pallet. With the exception of one heat exchanger, which became 4 degrees Celsius warmer than anticipated, the Environmental Control Subsystem kept all components performing as expected. Repressurising the airlock used 39 kilograms of oxygen from the Orbiter and 45% of the nitrogen supply available aboard Spacelab for this purpose.

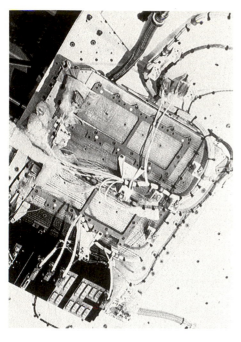

(Above) At the top is the rogue Remote Acquisition Unit, RAU 21, mounted on a freon-cooled cold plate.

(Below) The High Data Rate Recorder (HDRR), restored to operation by a Mission Specialist during the flight.

The Electrical Power Distribution Subsystem (EPDS) is not only essential to the operation of Spacelab subsystems but also constitutes a possible hazard. During the first Spacelab flight, data from a hundred and thirty-three standard Spacelab operational sensors and an additional twenty strategically placed components of the Verification Flight Instrumentation (VFI) were used to prove the EPDS. Readings were obtained from voltage- and current-measuring devices placed at various points including the principal elements of the subsystem and particularly the power distribution network. Voltages and currents were measured under normal working conditions, during varying loads and under different induced environments. In particular, the electrical surges introduced by switching on and off were measured. All measurements fell within their predicted ranges; thus the EPDS performed most effectively.

The Command and Data Management Subsystem (CDMS) is concerned with the communication links and with the control and display devices. Tests were designed to evaluate such functions as system activation, computer operation, software reconfiguration, Remote Acquisition Unit (RAU) operation, display capability and quality, Spacelab–Orbiter–ground data links and crew access to the CDMS. Almost three hundred operations or observations, some involving carrying out scientific experiments, were planned to prove the CDMS. This extremely complex subsystem worked well during the mission and was praised by the crew. There were, however, three faults, the last two of which were particularly significant. These were the only two major problems experienced during the entire first Spacelab mission.

The first fault occurred when a subsystem computer aboard Spacelab received two 'invalid' commands – and rejected them as such. The incorrect commands were traced to a computer aboard the Orbiter.

The second fault occurred 9 hours and 10 minutes after lift-off. The data acquisition functions of one of the fourteen Remote Acquisition Units, namely RAU 21, failed. This happened when the freon-cooled cold plate on which it was mounted became warmer than 22 degrees Celsius; however, another RAU mounted on the same cold plate was fortunately unaffected. RAU 21 was the main link with the NASA pallet-mounted experiments. The electrical power to these experiments was reduced so that less heat was transferred to the cold plate. Meanwhile, on the ground, the Experiment Computer Operating System (ECOS) was revised to overcome the problem and transmitted to the Orbiter–Spacelab complex. But there was to be a further complication. Some 50 hours later, the revised ECOS program 'crashed' due to the software modification introduced to overcome the problem with RAU 21. The computer was inoperative for 45 minutes. However, the computer was reloaded from the on-board Mass Memory Unit, and worked well for the remainder of the mission.

Considering all the experiments and the entire duration of the flight, nineteen different improvisations of computer programs were transmitted from the ground to the Orbiter, and twelve of these were implemented. The crew themselves also devised modifications to programs. Such changes carried out during the mission itself demonstrate yet another advantage of having manned vehicles in space.

The third fault occurred 86 hours into the mission. The High

Data Rate Recorder (HDRR) magnetic tape drive experienced excessive drag. This caused too large an electric current to flow in the tape drive motor. The tape recorder was automatically switched off. After opening up the recorder, Robert Parker, the Mission Specialist, rotated a stiff capstan by hand. This freed the blockage of the drive capstan. Thereafter the HDRR worked well for the remainder of the first Spacelab flight. In the interim 11-hour period, low-rate data were stored on a tape recorder in the Orbiter so that only a small amount of data was lost. Loss of the HDRR would have been serious, but the manual repair saved the scientific part of the mission.

A tape recorder aboard the Orbiter was also useful, in the programme of Verification Flight Instrumentation (VFI), when the special VFI recorder failed to operate for 9 minutes during the descent phase. This was the only significant problem experienced in the VFI tests.

The principal aim of the investigation of the Spacelab surface material was to demonstrate that the substances used for its exterior surfaces did not degrade in the space environment. For instance, changes in a surface's optical properties might affect astronomical observations. Also, any variation in the absorption of solar radiation would have to be taken into account in calculating the thermal conditions of on-board equipment. An array of thermal control coatings, containing suitable non-metallic materials, typical of those used on Spacelab, was mounted on the pallet. Pre- and postflight examination of the materials, together with temperatures observed by sixteen platinum resistance thermometers, permitted important conclusions to be drawn regarding the estimation of the thermal environment during future flights. The properties of the coating materials were deemed to be acceptable.

Viewed through the module end cone viewport, the IECM mounted on the pallet can be seen on the left. Alongside it are the helmet-shaped Atmospheric Emission Photometric Imaging (AEPI) instrument and the box-shaped solar spectrometer–radiometer.

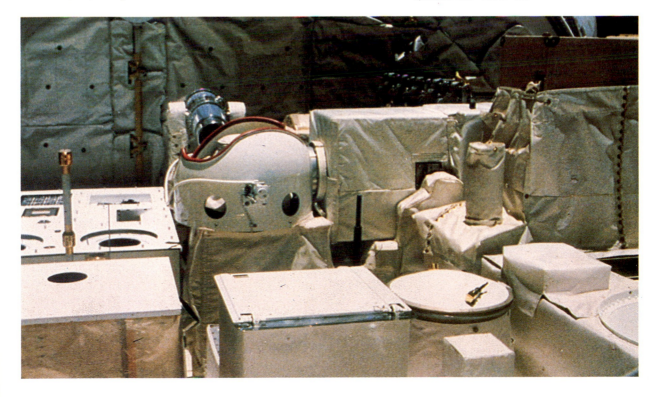

The Spacelab module has ample room for the crew to carry out the experiments. This is well illustrated in this view which shows Owen Garriott 'sitting down', reading his instructions for the next experiment. In the background is Byron Lichtenberg operating the Fluid Physics Module.

The objective of that part of the Verification Flight Test (VFT) programme concerned with the environment outside Spacelab was to ensure that the previously established contamination limits had not been exceeded. For this, a special Induced Environment Contamination Monitor, with a mass of about 350 kilograms, was designed. With supporting subsystems for power and data recording housed within the same container, this consists of ten monitoring devices such as a mass spectrometer for identifying contaminant species, a photometer for measuring the background optical intensity, and two micro-balances to determine absorption and desorption of contaminants in the Orbiter cargo bay. During the first Spacelab flight the IECM was attached to the pallet; it also flew on STS-2, STS-3 and STS-4. It will also be carried on later Spacelab flights.

## Some lessons learned

The first flight of Spacelab was thoroughly planned and time-line events were simulated to ensure that the best possible results would be obtained. Inevitably, however, not everything worked exactly as planned. Much has already been learned from both the ground and the flight phases of the mission; the techniques and procedures associated with a mission will be refined accordingly. Also, since many of the methods used were being applied for the first time, they can only be improved and made more efficient.

Most of the data acquired from the mission confirmed the effectiveness of the procedures adopted. In general, the team approach used was found to be very successful. This team-work continued from the early face-to-face discussions between the experimenters and the management team, through the thorough training of the Payload Specialists, into the integration phase. The Investigators' Working Group meetings provided an excellent preflight forum for the exchange of ideas amongst all team members. Further, the team spirit and efficiency were enhanced by realistic simulations that involved all the Payload Crew. During the flight itself, the Payload Operations Control Center (POCC) concept was shown to be very successful. The direct voice and video contacts between the Spacelab crew and the experimenter teams improved the quality of the data. The presence of scientists on-board proved to be invaluable. In this context, a better intercom system has been suggested by the crew, permitting three-way conversations between persons aboard Spacelab, in the Orbiter and on the ground.

The time-line, which schedules different tasks throughout the flight, can be improved. In particular, during the first two or three days of a flight, when the astronauts are becoming accustomed to the microgravity environment, the scheduling of events should not be tight and complicated tasks should be avoided. An adequate reserve of time is required to allow for unforeseen events and so that there can be a complete recovery of the schedule in later parts of the time-line.

Improvements in the operational control of some experiments can certainly be envisaged. Because, at present, three-quarters of the computer memory is occupied by operating procedures, a larger memory would be advantageous. Further, decentralisation by the use of dedicated microprocessors rather than the Spacelab Command and Data Management Subsystem for some experiments, should be employed wherever possible. This approach would not only simplify the software integration, but would also provide more operational flexibility. The RAU 21 fault illustrated that a single route for the flow of important data should be avoided. Two, or even three, units in parallel with each other would provide a 'fail-safe' data channel. Alternatively, on-board spares could be carried and fitted when necessary.

## Science around the clock around the world

The first Spacelab mission consisted of seventy-one experiments: fifty-eight from Europe, twelve from the USA and one from Japan (see Appendix). This is the greatest number of experiments ever conducted on a single spaceflight; thus Spacelab one was the most ambitious international mission yet conducted. The mass of equipment for European experiments exceeded that on all earlier ESA satellites. Of the seventy-one experiments, only two in Material Sciences and a major part of an Earth Observations experiment were unsuccessful. Altogether, fifty-one experiments were totally successful in that all their main objectives were achieved. Many, but not all, of the main objectives of the other seventeen experiments were accomplished. Overall, this is a remarkable success rate for a brand new laboratory.

These experiments, across a wide range of scientific disciplines, were performed in 10 days using thirty-eight instruments: sixteen were mounted on the pallet, twenty within the module, and the remaining two had components both on the pallet and in the module. The Material Sciences Double Rack was used for thirty experiments from eight European nations; the Fluid Physics Module was required for seven experiments from five European nations. Five of the US Life Sciences experiments were conducted in the NASA Minilab.

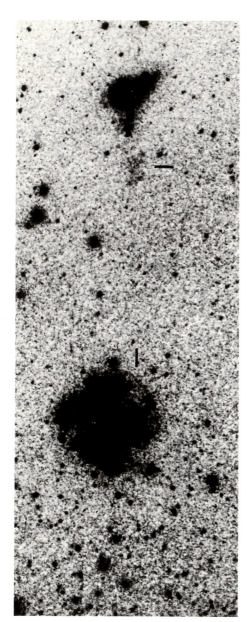

A negative enlargement of an ultra-violet image, observed by the Very Wide Field Camera, of the Large Magellanic Cloud (bottom) and the Small Magellanic Cloud (top). The horizontal bar indicates a region of hot stars bridging the gap between the two Magellanic Clouds. The vertical bar shows a thin arc-shaped feature in the Large Magellanic Cloud.

## Astronomy and Solar Physics

Observations were carried out in the ultra-violet and X-radiation parts of the electromagnetic spectrum to make both general surveys and detailed studies of special astronomical objects. Although Spacelab is not the ideal platform for conducting astrophysical research due to its relatively brief flight in space, three experiments (two ESA, one NASA) were performed and produced important results.

The Very Wide Field Camera, with a 60-degree field of view, was exposed to space using the airlock; it operated perfectly on days 6 and 7. Although the long twilight hours and the short nights caused the amount of data to be less than hoped for, forty-eight high-quality photographs in the ultra-violet were obtained of stellar clouds during the middle of the night. These photographs do not appear to have been contaminated by any Shuttle glow.

The Far Ultra-violet Space Telescope (FAUST), with a 7.5-degree field of view, should have observed faint sources of ultra-violet radiation from a variety of astronomical objects such as quasars and clusters of galaxies. Observations were made at wavelengths between 130 and 180 nanometres, but most of the photographs were overexposed and the film fogged. The light responsible for this could have been due to airglow emissions (at 130 and 136 nanometres) from oxygen atoms near 300 kilometres altitude in tropical regions (called tropical arcs), to Shuttle glow or to operation of the Orbiter's thrusters.

Some 200 hours of high-quality X-radiation astronomical data were obtained during the mission. The background level in space due to cosmic rays was found to be surprisingly low, so that the sensitivity of the X-ray spectrometer was greater than expected. Energy spectra have been obtained and celestial sources observed. Of particular interest are observations of Cygnus X-3, a galactic X-radiation source. A line in the spectrum was observed at 6.4 kilovolts, due to fluorescent radiation of iron. The amount of energy at this spectral peak varied by two to one cyclically over 4.8 hours. This is explained by the rotation of a very dense object, surrounded by its accretion disc, once every 4.8 hours.

In addition, three instruments (two ESA, one NASA) were pointed at the Sun to measure its energy output. Two independent, yet intercalibrated, measurements of the so-called solar constant were made with active cavity radiometers and with a dual-channel, self-calibrating radiometer. The latter instrument operated towards the end of the mission in two 6-hour periods of continuous sunlight. Its temperature was within the expected range – it could even have operated continuously. The accuracy of solar pointing of the Orbiter–Spacelab combination was very good. The enormous value of instrument calibrations both before and after flight in space has been well demonstrated by this experiment. However, a comparison between the two solar constant measurements was hampered by the RAU 21 problem. With the third solar instrument, spectral measurements were made at wavelengths where the Sun's output varies significantly with time, in the ultra-violet part of the spectrum, following calibration in space. Observations planned for later missions will lead to a study of the long-term variations of solar ultra-violet radiation.

## Space Plasma Physics

On the first Spacelab mission, the Earth's plasma environment, that is, the electrically charged gas in the uppermost atmosphere, was investigated by both passive and active means. Six experiments (four ESA, one US, one Japanese) were performed. High-intensity charged particle beams were injected into the space environment and the perturbations so produced were measured. In these interactive experiments space itself is used as a vast natural laboratory.

The SEPAC (Space Experiments with Particle Accelerators) equipment, provided by the University of Tokyo, was a particularly ambitious experiment. Regrettably, the high-power mode of the electron gun did not operate. The gun was designed to inject a high-power beam of electrons of up to 7.5 kilovolts energy with a current of up to 1.5 amperes in bursts of between 0.01 and 0.5 seconds' duration. The aim was to create artificial aurorae, to trace out geomagnetic field lines from one hemisphere to the other, and to stimulate plasma instabilities. A further purpose was to investigate the interaction between a low-power electron beam and the low-density, night-time ionospheric plasma. This experiment showed that the electric potential of the Orbiter–Spacelab duo took as long as 20 seconds to recover to its initial value after the electron gun had ceased operating. Other observations showed that ionospheric electrons were accelerated to energies several times greater than the electron beam energy in a beam plasma discharge. Details of the plasma processes responsible for this remain unknown, as are those which determine the response of the environment around Spacelab to the injection of beams of

The PICPAB detector is shown here outside the airlock as Columbia flies over the Sahara desert.

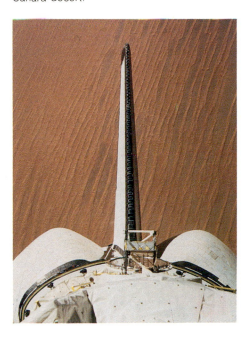

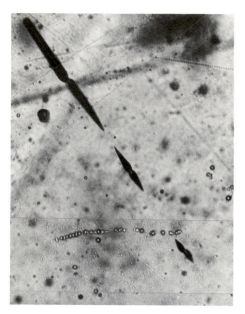

This photograph shows the last 1.6 millimetres of range of a cosmic-ray oxygen nucleus. Detected in a stack of plastic sheets, its existence was revealed by special chemical etching in the laboratory after the flight.

energetic electrons and ions. The low-power electron gun was operated simultaneously with the injection of either neutral nitrogen gas or argon plasma produced by an arc jet to investigate neutralisation of the beam. When the arc jet was operated alone, its excitation of airglow was measured by a photometer, a low light level, ultrasensitive television camera and also a NASA experiment (AEPI).

Studies of Phenomena Induced by Charged Particle Beams (PICPAB) was an ESA experiment with similar objectives. Using this, further investigations were devised after the partial SEPAC failure. PICPAB consists of a 7-to-10-kilovolt electron accelerator, with a beam current adjustable from 0.01 to 0.1 amperes, plus a 7-kilovolt hydrogen ion gun. On the second day of the mission, the gas pressure in this ion gun fell and the gun was rendered inoperable. Mounted in space in the Spacelab airlock, as far as possible from sources of interference, were several plasma diagnostic instruments and wave receivers sensitive over a wide frequency range from zero to 100 megahertz. When the electron gun stopped firing, it took a time varying from a fraction of a second to many seconds for the electric charge built up on the Orbiter to become neutralised.

A further experiment successfully measured the strength and direction of the geomagnetic field about which the energetic electrons spiral. Another observed the flux and energy spectrum of electrons, of between 0.1 and 12 kilovolts, whether of natural origin in the van Allen belts or produced by the electron guns. A detector consisting of a stack of plastic sheets recorded the tracks of cosmic rays with energies of millions of volts, heavier than the dominant hydrogen and helium nuclei; their identity and energy will be revealed by postflight analysis.

The Atmospheric Emission Photometric Imaging (AEPI) experiment itself produced some fine results. Magnesium ions were observed, at a wavelength of 280 nanometres, to be aligned along the geomagnetic field over sunlit tropical regions. These ions are produced in the ionosphere from disintegrating meteoroids.

A late addition to this experiment was carried aboard Spacelab to investigate further the Shuttle glow. This was a camera with a 300 lines per millimetre diffraction grating mounted in front of the lens to produce both a photographic image and a spectrum. The Shuttle glow, strongest near surfaces such as the tail when they are almost perpendicular to the Orbiter's direction of motion, occurs at wavelengths between about 450 and 780 nanometres. A line of molecular oxygen in the upper atmosphere in the infra-red, at 761.9 nanometres, is bright, as are a series of nearby hydroxyl lines. The Shuttle glow could be due to hydroxyl emissions, arising from an interaction between water absorbed on the Orbiter's surface and oxygen atoms that have gained energy by collision with the Orbiter which is moving through the oxygen atmosphere at 8 kilometres per second. If this were the explanation, however, more infra-red radiation would be expected at even longer wavelengths, beyond 0.78 micrometres, even to 1.5 micrometres. Such long-wave infra-red radiation was not observed. Thus, the hydroxyl emission hypothesis does not provide the full explanation for the Shuttle glow. This glow could also involve ionisation of the region near the Shuttle surface by a plasma instability.

Two methods are being suggested to overcome the problem of Shuttle glow for future missions making observations in the ultra-violet, visible and infra-red regions. One is to put the Orbiter into an orbit 200 kilometres higher than that used at present, where there are far fewer oxygen atoms. The other is to mount instruments such that they view directly out of the cargo bay and to orientate the Orbiter so that its underside moves first into the tenuous atmosphere.

## The Earth's atmosphere

In the Atmospheric Physics discipline, four experiments (three ESA, one NASA) were performed using sophisticated elevation scanning techniques as well as spectrometric imaging. The experimental techniques were very demanding because of the low light intensities available and because of the necessity for high spectral resolution. The Grille Spectrometer located on the pallet was designed to examine the atmosphere by analysing infra-red radiation coming from the Sun through the atmosphere. A complementary pallet experiment provided by NASA, the Imaging Spectrometric Observatory, measured the airglow spectrum at wavelengths ranging from the extreme ultra-violet to the infra-red.

The high-resolution Grille Spectrometer, operating in the infra-red wavelengths, suffered from an unforeseen computer problem. The Orbiter's Guidance, Navigation and Control computer was found to be set one day ahead and so the orbit calculations required to specify the instrument's viewing direction at sunrise and sunset were incorrect. This error was corrected by modifying the software on board. High-quality data were obtained until the supply of cooling gas was exhausted towards the end of the sixth day of the mission. Trace species such as ozone, carbon dioxide, oxides of nitrogen, water vapour and methane have been observed. The variation of the density of these important species with height, through the stratosphere and mesosphere, has been deduced.

The Imaging Spectrometric Observatory consisted of five spectrometers covering the spectral range from 30 to 1200 nanometres and observing the spectral signatures of many atmospheric species. Good dayglow emission data were obtained, although the problem with RAU 21 limited the amount of data produced. Further, the Shuttle glow was observed. The results obtained suggest that its cause is a combination of a plasma discharge and catalytic effects occurring on the Orbiter surfaces.

An infra-red camera with image intensification was used to observe large-scale (1000 kilometre) patterns of the region near 85 kilometres altitude where hydroxyl emissions naturally occur at night. Data were obtained between days 4 and 8; earlier, in-flight calibrations had been performed.

Lyman-alpha emission from atmospheric hydrogen was observed in the ultra-violet region at a wavelength of 121.6 nanometres, and from deuterium ('heavy hydrogen', with a nucleus comprising a proton and a neutron) at a wavelength of 121.5 nanometres. The density of deuterium at an altitude of 110 kilometres in the thermosphere was obtained for the first time ever during the first 2 days of the mission. Auroral emissions in the ultra-violet were observed on both the night and day sides of the Earth. On the tenth day of the mission, bonus results were obtained on the interplanetary medium and on two sources of 'hot hydrogen' emission. One was a star and the other a localised region of the Earth's atmosphere.

(Above) Photograph of infra-red emission from the night sky taken at the European Southern Observatory, La Silla, Chile, on 1 November 1975. These emissions originate from hydroxyl radicals near 85 kilometres altitude. During the 55-minute exposure, the Earth rotates causing the images of the stars to appear as arcs, known as star trails.

(Below) Observations of hydroxyl emissions made from Earth orbit give information over a large area and avoid the many difficulties of viewing them from the ground. This picture, obtained during the first Spacelab flight, shows the bright edge of the emitting layer; the streaks reveal wave structures in the upper atmosphere. A few stars are visible above the bright edge.

(Above) Ulf Merbold prepares the Metric Camera for use. This is a standard aerial survey camera modified for use aboard Spacelab.

# Earth Observations

Earth Observations were carried out with two instruments. The Metric Camera was used to take photographs covering 190 kilometres by 190 kilometres with high resolution (about 20 metres) of the Earth's surface on films measuring 23 centimetres by 23 centimetres. The crew mounted the camera in the optical window of the Spacelab module, changing the filters and magazines as necessary. On the third day of the mission, five hundred images were obtained on infra-red-sensitive film. Then a second magazine containing black-and-white film was inserted. After making twenty exposures, this film jammed. A scheme was devised on the ground to improvise a darkroom in a scientist–astronaut's sleeping bag. It was successfully used by Robert Parker to release the faulty magazine. With this, over five hundred

A false-colour infra-red photograph of the Red Sea coastline of Saudi Arabia, near Medina, taken with the Metric Camera on 2 December 1983.

Le Havre, France, was photographed from Spacelab by the Metric Camera, on 2 December 1983. The River Seine winds through verdant countryside which appears red in this false-colour infra-red picture.

Another Metric Camera photograph taken in the infra-red is of Gezira on the White Nile, Sudan. The thin red strips indicate irrigated areas growing millet and cotton.

(Right) Marseille and the mouth of the River Rhône, France, are photographed in black and white by Spacelab's Metric Camera. Sediment discharge by the River Rhône can be seen in the coastal waters.

(Below) A view from Spacelab one of the German–Austrian border region showing Munich (just above the centre), Innsbruck (lower left) and the mountains of the Tyrol, taken using the Metric Camera.

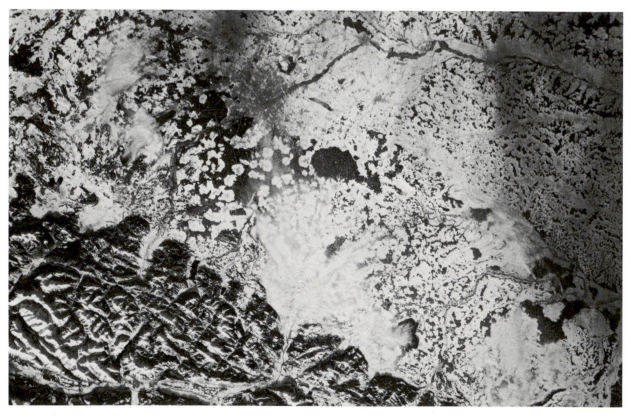

exposures of very high quality were obtained up to the ninth day. Fortunately the weather was good over most of the northern hemisphere target areas chosen, with few clouds obscuring the Earth's surface. However, the low solar elevation angles during November and December caused long shadows to be cast there, so that conditions were not ideal for photography over Europe and North America. An area of 11 million square kilometres (about 5% of the Earth's land surface) was photographed. These images are to be used for making detailed topographic and thematic maps at scales of 1:50 000 to 1:200 000. They will be used by more than a hundred groups around the world, many of which are in the developing countries.

The Microwave Remote Sensing Instrument should have used both active and passive microwave techniques near 10 gigahertz to observe the Earth's surface at all times. (This is possible since clouds are transparent to microwaves.) The instrument, mounted on the pallet, could be a forerunner of an all-weather remote-sensing system. However, the high-power amplifier of the transmitter required for the active mode failed inexplicably.

Three operational modes were to have been used, namely a two-frequency scatterometer, a Synthetic Aperture Radar (SAR), and a passive radiometer. In the event, only the passive radiometer mode provided results. From these the sea surface temperature below the Orbiter can be derived and a map of this built up on successive orbits. Because of the heat that the oceans transfer polewards, this information is of considerable importance in obtaining a better understanding of the Earth's climate.

A black-and-white photograph taken by the Metric Camera of the Black Hills of South Dakota, USA.

## Material Sciences in microgravity

Most of the Material Sciences experiments were housed in the Material Sciences Double Rack (MSDR). Crystal growth, fluid physics and metallurgy, all using a wide variety of specimens, were studied under microgravity conditions.

The MSDR, assembled in the Federal Republic of Germany, provides four major multi-user experimental facilities and hardware for three other experiments. These four facilities comprise three furnaces with different characteristics and a Fluid Physics Module. In addition it has its own central console for controlling and monitoring the equipment, a system which either provides inert gases (helium and argon) to the process chambers or evacuates them, an air and water cooling system, and a stabilised electrical power supply. This set of complex equipment operated well throughout the first Spacelab mission. The crew were essential for operating it, for exchanging specimens and for making a variety of visual observations.

The first of the three furnaces is the so-called Gradient Heating Furnace, provided by France. Equipped with three independent electrical heaters and a water-cooled rim, this provides heating up to 1200 degrees Celsius, with a controllable temperature gradient up to 100 degrees Celsius per centimetre. Three cylindrical cartridges, 27.5 centimetres long and 2.5 centimetres in diameter, containing the samples, can be mounted side by side in the furnace. When evacuated, this cools to room temperature in 20 hours; this time is reduced to 4 hours if helium is added when the furnace has cooled to 800 degrees Celsius. During the first Spacelab flight all fifteen

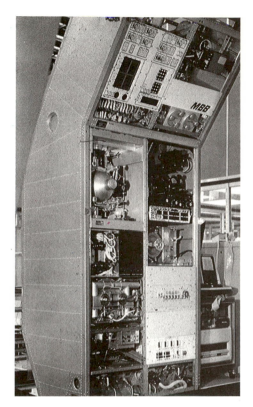

(Above) The Material Sciences Double Rack contains three multi-purpose furnaces and the Fluid Physics Module, as well as hardware (at the top right) for three other investigations. Altogether thirty experiments were conducted using the MSDR. These experiments could be controlled from the console (upper left), which was also used for monitoring their progress.

(Below) The type of cartridge used to place a sample in the Isothermal Heating Furnace.

cartridges were correctly treated under microgravity conditions. Special alloys of aluminium and zinc, aluminium and copper, and silver and germanium were processed. Alloys of tin were made to study the Soret effect. Crystals of lead telluride were grown and needles of tellurium-doped indium antimonide and nickel antimonide, of considerable interest to the semiconductor industry.

The second furnace is the Isothermal Heating Furnace, provided by the Federal Republic of Germany. It consists of two chambers, one for heating and one for cooling; each gives a constant temperature zone. One of these is heated under vacuum to any desired temperature between 200 and 1600 degrees Celsius and the other is kept at room temperature. The chambers can be moved back and rotated by 180 degrees so that the hot chamber covers the cold sample and the cold chamber envelops the hot sample. Thus the cold sample heats up as the hot sample cools down. The samples are contained in cartridges 10 centimetres long and 4 centimetres in diameter. This furnace experienced some initial problems during flight; a leaking chamber flange had to be replaced by a crew member. The furnace performed well until the ninth sample became stuck in the chamber because it had become deformed. This sample was then removed by the crew. After twelve of the twenty-two samples had been processed, the power supply for the heater failed, but only two experimenters did not obtain any samples treated in space.

Before the heating failure, samples were made of immiscible alloys of aluminium and indium, aluminium and lead, and zinc and lead. On cooling the molten mixture, heavy particles were distributed evenly in the light material. Unusual composite materials were also produced. Examples are short dendritic fibres, evenly distributed within a molten metal that is then solidified, and silicon carbide fibres in aluminium, a high-strength yet low-density material. The effects of small quantities of sulphur in liquid and solidified cast iron were examined, as were nucleation processes in silver–germanium alloys. Diffusion processes were investigated in molten glass. The behaviour of minute bubbles was examined during the melting and solidification of various metals and two-phase alloys. Such sponge-like alloys are of low density, yet strong, and have several practical applications.

The third furnace is the Mirror Heating Furnace, also provided by the Federal Republic of Germany. This has two ellipsoidal mirrors, at the common focus of which is the small sample (up to 11 centimetres in length and up to 2 centimetres in diameter); at each of the other two foci is a 400-watt tungsten halogen lamp. The sample can be rotated at between one-tenth and twenty revolutions per minute to attain a uniform temperature which can be adjusted to be between 200 and 2100 degrees Celsius. The furnace may be operated under vacuum or in an inert gas atmosphere. It can be rapidly cooled by flooding it with helium gas. This furnace operated from day 6 of the mission, but with an apparently reduced flow of cooling water so that it tended to overheat. One of the water pumps failed on day 9.

Each of the four investigators had one sample processed, nonetheless. A large single crystal of silicon of high quality was produced in space for the first time – such pure silicon crystals are in great demand in the semiconductor industry. This crystal, produced by the floating zone technique, is being examined for any inhomogeneities which could only be due to Marangoni convection. Single crystals of cadmium telluride and gallium antimonide, also required for semiconductors, were produced too. Spheres of liquid silicon, at 1410 degrees Celsius, were cooled and solidified. Since no convective disturbances can be present apart from those due to Marangoni convection, its effect can be studied in inhomogeneities and striations in the crystals that were grown in space.

The Material Sciences Double Rack also contains hardware for three other experiments. One piece of equipment, called the High Temperature Thermostat, consists of eight small furnaces which can either be isothermal (up to 1400 degrees Celsius) or have a temperature gradient applied to the specimen in the cartridge. This was used to investigate – in eight samples of liquid tin processed at different temperatures – the diffusion of the two stable isotopes of tin having different atomic masses (112 and 124 atomic mass units). Another piece of equipment is an Ultra High Vacuum (UHV) chamber capable of achieving a pressure less than $10^{-12}$ times atmospheric pressure. The forces of adhesion between a 0.3-centimetre-diameter metallic sphere and a flat metallic plate were measured by a piezoelectric transducer as the sphere repeatedly collided with the plate. Contamination on the metal surfaces was avoided in the vacuum conditions. The third piece is a low temperature chamber (−10 to 20 degrees Celsius) and called the Cryostat. Here, large single crystals of proteins such as lysozyme and beta-galactosidase were slowly grown. In fact, protein crystals were grown to from thirty to one thousand times their usual (Earth) sizes. With such large crystals, X-ray analysis can determine the molecular structure of enzyme proteins.

A silicon crystal grown in the Mirror Heating Furnace under microgravity conditions during the flight of the first Spacelab.

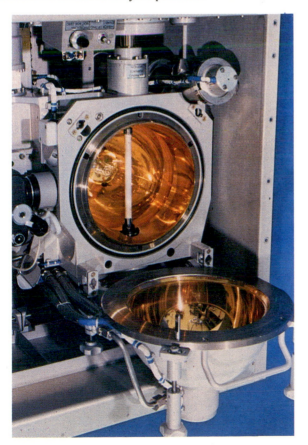

In this view of the Mirror Heating Furnace, with its door open, the rod of sample material is held vertically. The camera seen on the left can photograph the sample during processing in space.

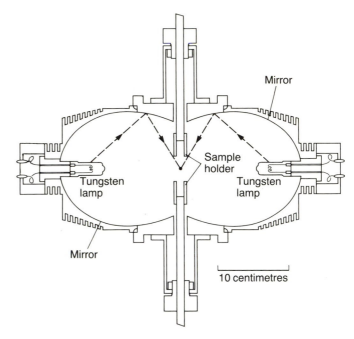

This diagram shows the Mirror Heating Furnace in cross-section. The ellipsoidal mirrors focus radiant energy from the halogen lamps onto the sample. The sample remains fixed and the rest of the furnace moves vertically, causing the melting zone to move along the rod.

A further three ESA experiments in the Material Sciences discipline used special equipment mounted outside the MSDR but within the Spacelab module. Of these, two were concerned with the growth from solution of organic crystals of exceptional quality, with high electrical conductivity (like metals), and of calcium or lead phosphate crystal platelets. The third, carried out as planned between days 2 and 8, was also a crystal growth experiment, but from the vapour phase. Mercury iodide was heated at one end of an airtight ampule and crystallised at the other, cooler end. Finally, one NASA experiment concerned tribological phenomena, namely frictional forces between two surfaces, under conditions of low gravity. The wetting and spreading of lubricants both on stationary surfaces and on moving ones such as bearings, were observed by ciné photography.

Made in Italy, the Fluid Physics Module consists of an airtight container supporting two similar discs, with diameters between 4 and 10 centimetres. Using this, the fundamental behaviour of fluids can be studied under microgravity conditions. A particular example of this is the physics of the 'floating zone', important in relation to the growth of crystals from the melt. A transparent liquid can be injected through one disc to produce a comparatively large volume of liquid held between the two discs. The discs can be rotated separately, at the same speed or at different speeds up to 100 revolutions per minute, in either direction. One disc can be either heated or vibrated axially at a chosen frequency up to 2 hertz. A potential difference of up to 100 volts can also be applied across the two discs. The size and shape of the floating zone of liquid formed between the discs, and motions within it, are observed by the crew and also filmed using two ciné cameras. The fluid physicists derived much benefit from the real-time television pictures of their experiments operating aboard Spacelab and from their conversations with the Payload Specialists.

The fact that Marangoni convection occurs was demonstrated for the first time on the first Spacelab mission. The behaviour of a column of oil that bridged the gap between two metal plates was studied when one plate was heated. This showed unambiguously that a Marangoni convection current was then set up in the liquid. In space, the length of the oil column exceeded 8 centimetres whereas on Earth the maximum length is only a few millimetres.

The shape of the outer 'free' surface of the liquid volume (known as a 'catenoid') is strongly influenced by surface tension forces and the geometry adopted. In one experiment, the shape was observed using time-lapse photography. The results will give new data on the weak van der Waals forces of attraction between the molecules of the solid and of the liquid. When the discs were moved apart, the volume of liquid split into two conical volumes on the discs plus a small drop between them. Such experiments also relate to the capillary behaviour of liquids in porous media.

The stability of the liquid volume was observed when it was stretched, vibrated or rotated. Liquids in differently shaped containers attached to a disc were vibrated and rotated, and both their internal motions and free surface (meniscus) shapes observed. These experiments are particularly relevant to the motions of liquids in containers in space, such as coolants or thruster propellants.

When an electric potential difference was applied between the plates and – by chance – a bubble of air had been introduced into the liquid, it was noticed that the bubble was conical rather than spherical. This observation suggests further experiments on electrophoresis at an air–liquid interface that should be conducted on later Spacelab missions.

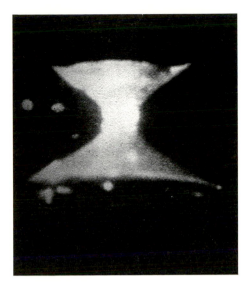

(Above) This column of silicon oil produced in the Fluid Physics Module was 8.8 centimetres long in space, whereas its maximum length on Earth would be only a few millimetres.

(Left) The Fluid Physics Module was used on Spacelab one for several experiments on the behaviour of fluids under microgravity conditions.

Byron Lichtenberg operates the Fluid Physics Module during the first Spacelab flight.

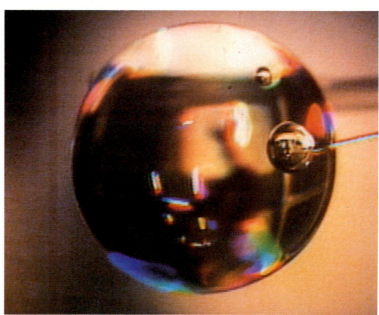

Using a hypodermic syringe (right), an air bubble is introduced into a liquid bubble floating in space.

## Life Sciences in microgravity

The Life Sciences investigations comprised fifteen experiments (eight ESA, seven NASA) designed to study the response of various biological systems to the conditions of microgravity and to energetic radiation – both electromagnetic and corpuscular – from space. The human vestibular system, located in the inner ear, controls both balance and the sense of position and orientation. Vestibular experiments were carried out which are the precursors to more detailed investigations to be performed using the Space Sled on later Spacelab missions. As well as providing important data for Earth-bound research, the results could have an important bearing on understanding space sickness. All the Mission Specialists and Payload Specialists participated in several medical experiments, not only during the flight itself but also before the flight and for 7 days immediately afterwards.

The so-called 'hop and drop' experiment involved measuring electrical activity in the calf muscle of one leg (electromyographic activity). For this, small metal discs were placed on the skin overlying the muscle. In an extension of this experiment, a hypodermic needle electrode was inserted just below and behind the knee to stimulate the muscular reflex (Hoffman response) – the 'drop and shock' experiment. The astronaut was accelerated towards the floor of the Spacelab module by forces applied using elastic cords, and bounced. The observations in these experiments, which relate posture and vestibular stimulation, were displayed to the crew members on a real-time oscilloscope as well as being recorded for later analysis.

The coupling between eye movements and vestibular reflexes was studied in various experiments. Small linear and angular accelerations were applied to a crew member. A simple seat and Body Restraint System (BRS) prevented him from touching the walls of Spacelab and also restrained his head movements. He wore a helmet which projected patterns onto one eye (a technique known as optokinetic stimulation). Movements of the other eye were observed using a high-resolution television camera. The astronaut also reported the size and direction of the movements that he perceived. It was found that in the near-weightless conditions on-orbit, the eye plays a most important part in providing information to the brain to orientate the body. Also it was discovered that the ocular reflex called 'nystagmus' could be produced after a few days in space. This is the characteristic twitching movement of the eye noted after the head rotates. It is actually a reflex movement maintaining visual fixation on a stationary object while the human body moves.

Another part of these experiments stimulated the inner ear by blowing cool air or warm air into the outer ear. Surprisingly it was found that this caused appreciable eye movements. These results tend to disprove the commonly held view that convection in the fluid of the inner ear is mainly responsible for nystagmus.

In another experiment, a scientist–astronaut viewed a rotating dome lined with a coloured 'polka-dot' pattern, as he either rotated freely or had his feet restrained. The positions of his eyes were observed using a video camera, and he noted his feelings concerning rotation. In a further part of this experiment, he was blindfolded and strapped to a flat surface. After 5 to 15 minutes rest, he was asked to point to marks on the Spacelab's walls and to describe the positions of his limbs. If the straps were rather loose, he was less certain of his orientation in the module. Additionally, on returning to Earth, a scientist–astronaut with his eyes closed experienced some difficulty standing upright. These results all indicate that the body's nervous system reinterprets signals from the otolith organs as it adapts to weightless conditions, and that readaptation to the one-$g$ conditions on the Earth's surface takes some days.

The Mission Specialists and Payload Specialists took turns wearing a light yet rigid backpack. This supported three small accelerometers, measuring accelerations in three mutually perpendicular directions, an electrocardiograph (ECG) observing heart activity, and a miniature four-track tape recorder recording the observations. The purpose of this study of ballistocardiography in three dimensions was to investigate movements and accelerations of the levitated human body caused by the heart's pumping of blood around the body's arteries. This was done when breathing normally, when

The scientist–astronauts were subjected to 'hop and drop' and 'drop and shock' tests using the equipment shown here.

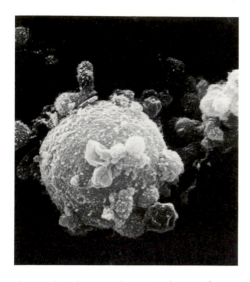

A scanning electron microscope image of a human lymphocyte flown on Spacelab one. The cell is 4.5 micrometres across.

holding breath and when squatting, both after 3 minutes of physical exercise and after resting.

Ulf Merbold wore a belt-mounted miniature magnetic tape recorder, of the type much used in clinical medicine, for most of the flight. On this were recorded time, brain activity (an electroencephalogram, EEG), eye movements (an electro-oculogram, EOG) and, during ascent and descent, heart activity (an electrocardiogram, ECG). The number of eye movements observed whilst the astronaut was sleeping was much higher than normal during his first night in space. The aim was to investigate the physiological reactions of a representative scientist, who is not a test pilot, to the unusual rigours of spaceflight.

The study of lymphocyte proliferation in space provides important information for cell biologists and immunologists. This is because lymphocytes, which constitute about 30% of the white blood cells, give rise to antibodies that combat infectious diseases. Studies in an incubator held at 37 degrees Celsius (body temperature) extended those carried out earlier aboard STS-8 and the Soviet Soyuz 6, 7 and 8 missions. They have shown that lymphocyte activation in the

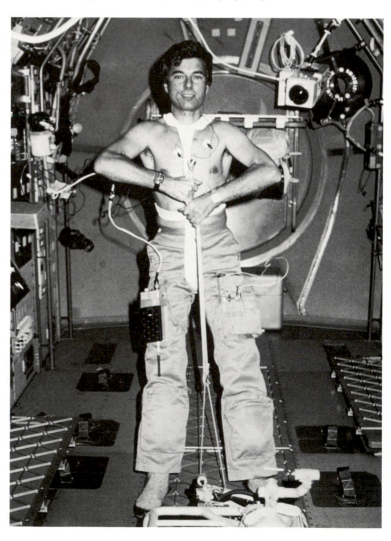

Ulf Merbold is exercising vigorously before floating freely in the module and having his ballistocardiogram recorded.

microgravity conditions of space is significantly less than in one-*g* conditions at the Earth's surface. This effect could determine the maximum safe duration of a space mission. This research is also valuable for ground-based medical research on the prevention of diseases.

Blood samples were taken from crew members throughout the flight, stored, and analysed for hormone levels after the flight. Studied in particular were prostaglandins, which might regulate the body's mineral metabolism on-orbit. The decrease in the amount of red blood cells (erythrocyte kinetics) in the early phases of the mission was demonstrated, and the simultaneous decrease in the volume of plasma (the colourless blood fluid) was measured. While taking some blood samples from a vein near the elbow, the pressure of blood returning to the heart was measured with a strain-gauge and related to the movement of body fluid from the lower part of the body to the upper part. This pressure was found to have decreased – an unexpected result.

A quite different type of microgravity experiment was concerned with the ability of an astronaut to distinguish between objects of different mass but otherwise identical. The objects used were twenty-four lead and epoxy resin balls, each of 3 centimetres diameter and labelled with a different letter, with masses between 50 and 64 grams in intervals of 2 grams. The astronaut moved each of a pair of balls, judged which was the more massive, and recorded this on a card. After about 20 minutes, he had done this for seventy-two pairs. Before the mission, about ten pairs were misjudged – in space this rose to around seventeen. For up to 2 days after returning to Earth, the astronauts did not judge mass well, but after 4 days their performances had returned to their preflight standards. It is concluded that scientist–astronauts can, with a probability of 0.75, correctly distinguish between two balls whose mass difference is 4.5 grams in the one-*g* condition at the Earth's surface. However, a greater mass difference (8.3 grams) is necessary in the microgravity environment of Spacelab. This result suggests that Man is more sensitive to weight than to inertial mass.

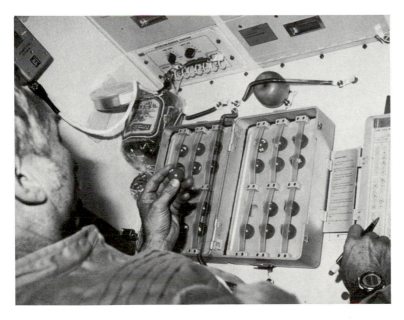

Owen Garriott tries his hand at distinguishing between apparently identical balls as his contribution to the mass discrimination experiment. Garriott's apple, unlike Newton's, has to be held down to prevent it from floating away in the microgravity conditions inside the Spacelab module.

Ulf Merbold acts as a gardener in space, tending his sunflower seedlings.

Dwarf sunflower seedlings (*Helianthus annuus*) were tended by Ulf Merbold; he also planted some sunflower seeds. These were all grown in the dark and, for one or two days during their development, observed automatically every 10 minutes using infra-red illumination and an infra-red-sensitive television camera. Although problems occurred with camera synchronisation, it was found that the tips of the seedlings traced out elliptical paths every $108 \pm 3$ minutes. This observation of nutation in the near-weightless conditions in Spacelab indicates that nutation arises from within the plant – the Darwinian view. It does not favour the idea of the plant having a servo mechanism which seeks – but does not accurately find – a preferred direction of growth, the vertical, as specified by the direction of the acceleration due to gravity.

Circadian rhythms, with a period of about one day, in the growth of the common fungus *Neurospora crassa* were observed in space, even though the fungus was kept in constant darkness. However, in space, the clarity of the rhythm was less than for samples similarly grown in Earth-bound laboratories.

Biological specimens and bacterial spores (*Bacillus subtilis*) exposed to the radiation and vacuum environment of space were returned to Earth for detailed analysis. Cosmic ray particles of high atomic number and high energy (hundreds of millions of volts per atomic mass unit) can damage or even destroy cells of living organisms. In the advanced Biostack experiment, layers of different biological organisms were sandwiched between both silver chloride and plastic detectors of cosmic rays. In this way, the trajectory of each cosmic ray particle could be localised in the biological layer and the damage that it caused specified. There were three Biostacks in the module and one on the pallet. Fifteen standard radiation dosimeters were also placed in the Spacelab module to map the radiation there due to cosmic rays and to the van Allen radiation belts. The results provided information on the radiation environment useful to future astronauts and experimenters aboard Spacelab orbiting the Earth at an inclination of 57 degrees. Although the level of radiation is low, it is interesting to note that for Spacelab one it was about twice that observed in the lower inclination (28.5 degrees) orbits of earlier STS flights.

## Epilogue

The great overall success of the scientific programme of the first Spacelab mission was undoubtedly due to the dedication of all members of the experimental and industrial teams and of all NASA and ESA staff involved. A very important contribution to its success came from members of the Spacelab crew themselves. Not only did they exhibit skill, they also demonstrated that they were fully prepared to deal with any situation that arose. They modified computer software and improved instrument settings. They repaired broken equipment and they improvised a photographic darkroom. In conjunction with experimenters on the ground, with whom they were in voice or video contact, they even devised new investigations. In fact, the team spirit between Man-on-the-ground and Man-in-space which was developed and clearly demonstrated in the first Spacelab mission points the way forward for future research in Earth orbit.

## Beyond the first Spacelab mission

Spacelab represents a European investment in the future of manned spaceflight. The first Spacelab mission has demonstrated a new concept – that of an international and multi-disciplinary laboratory in space. This space laboratory can not only be used for further Spacelab flights, but it can also be the foundation stone for more ambitious space plans. Building upon the experience of Spacelab, these will surely lead eventually to a permanently manned Space Station. Plans for the use of Spacelab over the next few years are already well advanced.

# ORBITING LABORATORIES OF THE FUTURE

*'We can follow our dreams to distant stars, living and working in space for peaceful, economic and scientific gain.'*
President Ronald Reagan

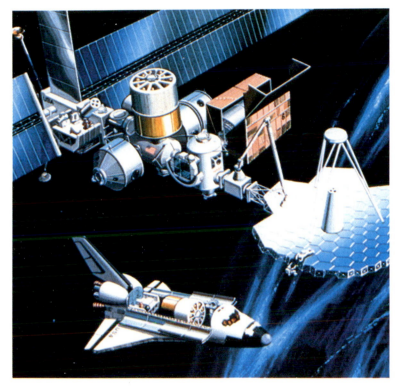

Modified Spacelab modules and pallets could become the building blocks of a manned Space Station, as illustrated in this artist's impression. Its main features are a large wing-like solar array producing many kilowatts of electrical power, a rectangular panel radiating waste heat, modules for people living and working in space and pallets supporting automated experiments. In this picture a large antenna is being constructed, and a Space Shuttle is arriving with workers and materials from Earth.

## The next four Spacelab missions

Paradoxically, Spacelab three will actually be launched before Spacelab two because of delays in building the complex Instrument Pointing System which will be used to orientate the four solar telescopes on the Spacelab two mission.

For ***Spacelab three*** the emphasis is being placed on microgravity experiments in Life Sciences and Material Sciences and Technology. For example, equipment to be flown on Spacelab four will be tested; mercuric iodide crystals will be grown from a vapour and triglycine sulphate crystals from solution. Other experiments range from Astrophysics to Atmospheric Observations and Fluid Physics. In all, fourteen different experiments with a mass of about 2500 kilograms will be performed.

## Planned Spacelab missions

| Date | Designation and Orbiter | Orbital inclination (degrees) | Spacelab configuration | Major mission discipline |
|---|---|---|---|---|
| January 1985 | Spacelab three SL-3 Discovery | 57 | Long module +a pallet-like structure | Microgravity (Material Sciences and Life Sciences) |
| April 1985 | Spacelab two SL-2 Challenger | 50 | Igloo +three pallets | Astronomy |
| October 1985 | German D-1 Columbia | 57 | Long module +a pallet-like structure | Microgravity |
| November 1985 | Earth Observation Mission EOM-1 Atlantis | 57 | Short module +one pallet | Atmospheric Physics and first Spacelab reflights |
| January 1986 | Spacelab four SL-4 Challenger | 28.5 | Long module | Life Sciences |
| May 1987 | Spacelab eight SL-8 Atlantis | 57 | Long module +one pallet | International Microgravity Laboratory |
| January 1988 | Japanese SL-J Challenger | 57 | Long module +one pallet | Microgravity |
| June 1988 | Spacelab ten SL-10 Columbia | 57 | Long module +one pallet | Life Sciences |
| November 1988 | German D-2 (or D-4) Atlantis | 57 | Long module +a pallet-like structure | Microgravity |

Plans for future space missions are always subject to change due to unforeseen circumstances. The data given here and on page 165 are based on information available in June 1984.

The inclination of the Spacelab three orbit will be 57 degrees, just as for Spacelab one; however, the altitude of the Orbiter with Spacelab in its cargo bay will be higher, at 370 kilometres. The orientation of the Orbiter–Spacelab combination will be maintained with the Orbiter's tail pointing towards the Earth. This orientation relies on gravity-gradient stabilisation, with the number of thruster firings being kept to a minimum. In this way, the best possible microgravity environment will be maintained inside Spacelab, maximising the chance of growing perfect crystals.

An experiment which relies on the microgravity conditions in Spacelab three concerns the dynamic behaviour of liquid drops. Photographic studies are to be made of drops that are spinning or vibrating. Such fundamental investigations are relevant to physics on the smallest scale, that is, to nuclear fission and fusion, and also to physics on the largest scale, such as the dynamics of stars. Another experiment in Fluid Physics simulates large-scale flows, waves and instabilities in the atmosphere of a rotating planet or star. The simulation is performed using an insulating fluid confined between two concentric, rotating, electrically conducting spherical shells. The two spheres are kept at different temperatures, with the dielectric constant of the fluid depending on temperature. When a potential difference (a voltage) is applied across the spheres, a force acts on electric charges in the dielectric. This force simulates the gravitational force acting on gas molecules in the atmosphere. The motion of dye injected into the fluid is to be recorded photographically and interpreted later.

There is a spectroscopic investigation of the absorption of infra-red radiation from the Sun, at wavelengths from 2 to 16 micrometres, during sunrise and sunset as seen by Spacelab. This enables the concentrations of trace gases in the stratosphere and mesosphere to be determined in order to find their variations with latitude, longitude and altitude.

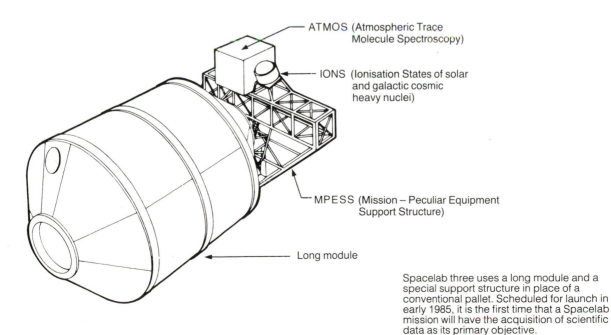

ATMOS (Atmospheric Trace Molecule Spectroscopy)

IONS (Ionisation States of solar and galactic cosmic heavy nuclei)

MPESS (Mission – Peculiar Equipment Support Structure)

Long module

Spacelab three uses a long module and a special support structure in place of a conventional pallet. Scheduled for launch in early 1985, it is the first time that a Spacelab mission will have the acquisition of scientific data as its primary objective.

An Indian experiment is to observe cosmic rays from the Sun and our Galaxy. Cosmic ray particles, multiply charged ions ranging from oxygen to iron, are to be measured at energies up to 100 million volts per atomic mass unit by a specially designed detector. Of 40 centimetres diameter, this consists of stacks of thin plastic sheets. The tracks of the cosmic ray particles are revealed by chemical treatment in the laboratory on return to terra firma.

An important late addition to this mission is the Very Wide Field Camera experiment. This experiment, to make a general survey of the celestial sphere in the ultra-violet region of the spectrum, was first performed on Spacelab one with promising results. A reflight was guaranteed by NASA because the delayed launch resulted in periods of total darkness that were too brief for the best results.

***Spacelab two*** will consist of the 'igloo', providing power, cooling and command for the three pallets which support eight instruments exposed to space, and a very large (1800-kilogram) cosmic ray detector. For this flight there will be no pressurised module. Thus the scientist–astronauts will view and control the pallet-mounted experiments from the aft flight deck of the Orbiter. The six-man crew will orbit the Earth at an altitude of about 400 kilometres for 7 days. The experiments have a total mass of 5000 kilograms. The investigations, ten originating from the USA and two from the United Kingdom, are in the disciplines of Astronomy, Solar Physics, Space Plasma Physics, Atmospheric Physics, Space Technology and Life Sciences.

The IPS is checked by engineers at Dornier Systems in Friedrichshaven in the Federal Republic of Germany. The instruments are attached to the large circular ring. The gimbal systems and drive mechanisms for the control in roll, elevation and azimuth, together with the associated electronics (covered by a heat shield during flight) and supporting framework, can be discerned.

The large detector of the cosmic ray experiment is being lowered before being attached to the pallets of Spacelab two during payload integration. The infra-red telescope with its protective cap is seen in the foreground.

(Below) This image of the constellation of Orion, a region where new stars are being formed, was prepared from data taken by the Infra-red Astronomy Satellite (IRAS). It illustrates the wealth of detail that can be obtained from observations in the infra-red region of the electromagnetic spectrum. Strong radiation at wavelengths near 100 micrometres is shown in red and yellow, with that at 12 micrometres in blue. The Horsehead Nebula and the Orion Nebula are the bright patches (lower centre).

The cosmic ray experiment will measure the composition and energy distribution of extremely energetic cosmic ray ions. The counters are housed in a dome-topped, pressurised cylinder 4 metres high and mounted at the rear of the payload bay. Forward of this is an infra-red telescope, cryogenically cooled with liquid helium ($-269$ degrees Celsius; 4 Kelvin). Complementing observations of compact astrophysical sources made by the very successful Infra-Red Astronomical Satellite (IRAS) of the USA, the United Kingdom and the Netherlands, the infra-red telescope on Spacelab two will study enormous extended sources of infra-red radiation in the Universe. It will also investigate the infra-red glow of the Orbiter itself.

The middle pallet carries two identical X-ray telescopes designed to produce images of clusters of galaxies and other extended X-radiation sources. Over 3 metres long, this British experiment with its own pointing system will obtain high-resolution pictures of the sky at high energies. The spectral feature between 6 and 7 kilovolts that is associated with iron will be observed in detail.

The remaining pallet carries the 1265-kilogram Instrument Pointing System (IPS) which is being developed by Dornier Systems in the Federal Republic of Germany as part of the Spacelab Programme, and provided as a service for Spacelab users. The purpose of this highly sophisticated system is to ensure not only the accurate pointing but also the ability to hold that pointing for instruments (of up to 2000 kilograms) such as telescopes. By mounting such instruments on this system, they can be accurately aimed at a particular point in space, or on the Earth or on the Sun, with better accuracy than that provided by control of the orientation of the Orbiter vehicle itself ($\pm 0.1$ degrees). The IPS is also needed to correct for the appreciable movement between the Orbiter and an instrument situated on the pallet.

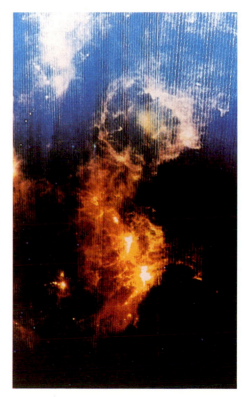

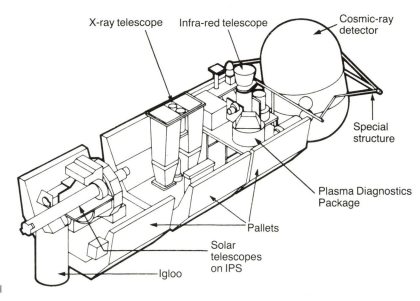

Spacelab two will use the pallet-only configuration, with the scientific objectives being mainly dedicated to Astronomy. It will see the first use of the ESA-provided Instrument Pointing System (IPS).

The emblem of Spacelab two indicates the scientific disciplines to be explored, emphasising the solar experiments, and depicts the Spacelab configuration chosen for the flight.

The IPS provides three-axis attitude control (along the 'roll' axis, that is, the line of sight of the experiment, and two axes perpendicular to this). The accuracy of pointing is about 5 arc seconds (just greater than one-thousandth of a degree) in the axes perpendicular to the line of sight and about 20 arc seconds in roll about the pointing axis. This is done by a three-axis gimbal system and reference to stars in known positions. The IPS is extremely sensitive to outside disturbances, particularly those introduced by the motion of crew members. When a crew member in the module pushes himself away from a wall to reach some other part of the module, he causes an additional inaccuracy of pointing of 4 arc seconds in the two perpendicular axes, and some 15 arc seconds in roll.

During ascent and descent, the IPS is stowed horizontally on the pallet. It is mechanically disconnected from its payload which is supported by the specially designed clamp. In this way, the IPS does not impose any loads onto the payload, or vice versa. Once in orbit, the payload (usually a telescope) is released from its clamp and mechanically connected to the gimbal system via a flat attachment ring. The IPS performs a 90-degree manoeuvre to an upright position. It is then operational.

Overall control of the IPS during normal operations is exercised from the Spacelab control centre via the subsystem computer. Tables of information on the reference stars and on the objects to be observed are an important element of the software which is resident in the Mass Memory Unit of the Command and Data Management Subsystem (CDMS). Thus, IPS operation is an optimal mix of complex automated functions and manual action and judgement. The IPS will first be used on this Spacelab mission to provide accurate pointing for the solar experiments investigating helium abundance, sunspots and related phenomena, and the structure of solar magnetic fields.

One of the solar experiments on Spacelab two is the British Coronal Helium Abundance Spacelab Experiment (CHASE). Using a new ultra-violet resonance scatter method, this will accurately measure the abundance of helium, relative to that of hydrogen, in the Sun's atmosphere – the corona. This value of the relative abundance of helium to hydrogen is believed to be representative of the interstellar gas at the time that the Sun was formed, and indeed of the entire Universe. As such it indicates the amount of helium produced in the Big Bang origin of the Universe. It is therefore an important experimental test of different cosmological models.

Another solar experiment investigates the strength of the magnetic field and the movement of solar plasma in the vicinity of plage regions and sunspots. This is done with high temporal resolution and good spatial resolution. Even the granule patterns will be investigated. Mechanisms whereby the chromosphere and corona are heated will be investigated in the third experiment using observations made with high spectral resolution and good spatial resolution.

A further experiment mounted on the IPS monitors the intensity and spectrum of solar ultra-violet radiation and its variability. Because of its absorption by the Earth's atmosphere, these results will provide important information on physical and chemical processes occurring in the stratosphere.

In the field of Space Plasma Physics, a Plasma Diagnostics Package (PDP) will operate in the cargo bay and around the Orbiter. It will be picked up from the pallet supporting the infra-red telescope by the RMS. The instrumented package will be released; this sub-satellite will then drift up to 10 kilometres behind the Orbiter. It will investigate the interaction of the Orbiter with the surrounding ionospheric plasma, that is, the wake of the Orbiter. The Orbiter will return to the PDP so that it may be recovered, again using the RMS arm, and replaced in the cargo bay.

The CHASE instrument is used to study the Sun's corona in the extreme ultra-violet region of the electromagnetic spectrum. The delicate arc second pointing accuracy required by the instrument will be satisfied by the IPS.

The IPS is carried into space on a pallet in the position depicted here. Long telescopes mounted on the IPS are supported by additional pallets during launch, re-entry and landing.

Another experiment involves firing the Orbital Maneuvering System (OMS) thrusters and investigating their effects on the ionospheric plasma. The plasma density is expected to decrease, electrons in the ionosphere recombining with positive ions formed from chemicals of the thrusters' exhaust gases. Such plasma depletions will be studied by ground-based radio and airglow instruments situated around the world. Radio and airglow observations will also be made by a research aircraft. The purpose of such investigations is to learn more about ionospheric processes. It may even be possible to create a preferred radio propagation path for ground-based very high frequency (VHF) studies in radioastronomy or for very low frequency (VLF) whistler-mode studies of the Earth's magnetosphere.

A third experiment on the PDP concerns electrical charging of the Orbiter. An electron gun emits a pulsed beam into space, so changing the Orbiter's electric potential with respect to that of the surrounding plasma. The return current is measured.

The Plasma Diagnostics Package is moved around in the vicinity of the Orbiter cargo bay during STS-3 by the Remote Manipulator System (RMS) arm. During Spacelab two the PDP will be released to become a free-flyer, referred to as a sub-satellite, and later recovered by the RMS arm.

The behaviour of liquid helium will be studied under microgravity conditions. At a temperature of 4 Kelvin this is termed a superfluid – it has most unusual thermal and viscous properties. Liquid helium has practical cryogenic applications in terms of superconducting magnets needed to produce intense magnetic fields.

One of the two experiments in Life Sciences will study the formation of lignin in plant seedlings in a controlled-oxygen, microgravity environment. This will follow up experiments carried out during STS-3. In a medical experiment, blood samples are to be taken from the astronauts before, during and after the flight. Changes in the astronauts' mineral metabolisms will be studied with the aim of understanding the cause of bone demineralisation in space. Loss of calcium in the bones could limit the duration of an individual astronaut's spaceflight.

The Federal Republic of Germany is sponsoring the microgravity Spacelab mission called **D-1**. This is the first Spacelab mission to be outside the control of NASA. Although NASA will provide the Shuttle and crew, as well as ground and flight support, the mission management rests with the D-1 management team at Köln-Porz in the Federal Republic of Germany. The Payload Operations Control Centre will be at Oberpfaffenhofen.

The Material Sciences Double Rack (MSDR) that was flown on Spacelab one will be refurbished and flown again on the D-1 mission. Crystals free from defects are to be manufactured, as are metal alloys with special properties. An advanced Fluid Physics Module and a Process Chamber will be installed for the study of fluids, as will a Materials Laboratory (MEDEA) to investigate the behaviour of materials at high temperature. Navigation experiments (NAVEX) will be performed with equipment placed outside the module on a simple structure which also supports a NASA-provided Material Experiment Assembly (MEA) containing a number of automatic facilities.

The European and US vestibular experiments on Spacelab one, concerning the effects of both accelerations on Man in space and stimuli applied to the eye and the inner ear, are to be extended during D-1. This will be done using the Space Sled, an ESA-provided facility for vestibular research. It consists of a seat running on rails which are fixed to the floor of the Spacelab module. The position of the seat can be changed so that an astronaut sitting in it faces forwards, sideways or upwards. He can then be accelerated, at up to $0.2\ g$, along the length of the module. An oscillating acceleration can also be applied and the subject's responses measured. In this experiment the astronaut, wearing a helmet which prevents him from seeing his surroundings, can only detect acceleration changes via the vestibular mechanisms of the inner ear. Extra visual information can be introduced to the Sled-seated astronaut by means of a small television monitor, mounted in the specially designed helmet, in front of one of his eyes. These images, including moving patterns, can provide conflicting information to the brain of the astronaut. His responses are then recorded by an infra-red television camera, also in the helmet, observing the other eye's reactions.

Another multipurpose facility aboard Spacelab for the D-1 mission is the ESA-provided Biorack. This is designed for a variety of biological investigations on plants, tissues, cells, bacteria and insects in the unique microgravity and radiation environments of space. The Biorack has two centrifuges providing a one-$g$ environment,

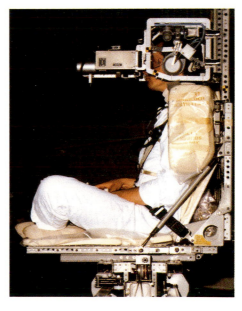

The ESA-developed Space Sled produces a controlled, known acceleration of a human test subject. This photograph shows a subject seated on the Sled wearing the helmet-like device for stimulating and measuring responses of the human vestibular system. The Sled will be flown for the first time on the D-1 flight.

The Payload Crew for the forthcoming German Spacelab mission, D-1, are seen during a break in their training. The NASA Mission Specialists are Guion Bluford (far right) and Bonnie Dunbar, and the three Payload Specialists are the Europeans Wubbo Ockels (far left), Ernst Messerschmid (third from left) and Reinhard Furrer (second from right). Ulf Merbold, ESA's first man in space, is the back-up Payload Specialist for this flight.

allowing data obtained under microgravity and one-*g* conditions to be compared, and two incubators holding up to fifty samples of 50 millilitres or twelve samples of 300 millilitres. The incubators are set to different temperatures, one near 20 degrees Celsius and the other near 40 degrees Celsius. A cooler (operating at 4 degrees Celsius) and a freezer (at −15 degrees Celsius) will be used to preserve up to twenty samples. Using a glove box system, the Payload Specialist will be able to handle biological cultures and specimens safely in a closed volume, through which filtered air is circulated. The specimens can be observed or photographed through a microscope.

About fourteen experiments will be performed with the Biorack on the D-1 mission. Most are concerned with cell structure, cell proliferation, cell functions and other aspects of cell behaviour. White blood cells, human lymphocytes, root tips of lentil seedlings, green algae and bacterial spores are all to be investigated, as are *Escherichia coli* bacteria and eggs of the much studied fruit fly *Drosophila*. The earliest stages of embryonic development of the amphibian *Xenopus*, which determine the future dorsal and ventral sides of the body, are to be studied. Further, the radiation environment inside the Spacelab incubators is to be recorded, as are the effects of this radiation on the eggs of stick insects.

The Payload Crew for the D-1 mission illustrates the international nature of Spacelab. The Mission Specialists will be Bonnie Dunbar and Guion Bluford of NASA. The three Payload Specialists will be Wubbo Ockels of ESA and two West Germans, Ernst Messerschmid and Reinhard Furrer. Although Dunbar will be on her first spaceflight, Bluford has already flown on STS-8. The three Payload Specialists are first-timers, but Ockels acted as back-up Payload Specialist for Spacelab one. The back-up role in D-1 will be fulfilled by ESA's first man in space, Ulf Merbold. The total crew of eight will be the largest Shuttle crew ever.

The ***Earth Observation Mission*** (EOM-1), the first to use a short module, has a strong international flavour. Its objectives are to study the Sun and the Earth's atmosphere using existing equipment that was developed for the OSS-1, Spacelab one and Spacelab three missions. Two of ESA's Spacelab one experiments for which the lighting conditions were unfavourable will be reflown. These are the Grille Spectrometer, which evaluates trace constituents in the atmosphere, and the Metric Camera to take pictures of the Earth for mapping purposes.

The Metric Camera which took many fine pictures from Spacelab one will fly again on EOM-1 when the lighting conditions over Europe should be better. The Metric Camera is shown here mounted on the optical window of the Spacelab module.

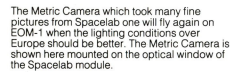

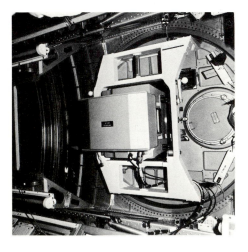

The other first Spacelab experiments are three solar experiments (1NS 008, 1ES 016 and 1ES 021, see Appendix) and the remaining four NASA pallet-mounted experiments. The latter are the Japanese active experiment (1NS 002) to probe the ionosphere and magnetosphere, the Far Ultra-violet Astronomy Space Telescope (FAUST, 1NS 005) and the instruments designed to investigate the Earth's environment, namely the low-light television equipment (1NS 003) and the Imaging Spectrometric Observatory (1NS 001). It is planned that these pallet experiments will be flown about once per year; they would then calibrate other similar instruments.

ESA's scientist–astronaut, Claude Nicollier, a Swiss citizen, will fly as Mission Specialist on EOM-1, together with Owen Garriott, the American Mission Specialist who flew on Spacelab one. The Payload Specialists, Byron Lichtenberg and Michael Lampton, have considerable experience with many of the experiments from their training for the first Spacelab flight.

## Future Spacelab missions

The **Spacelab four** mission will include a repeat of Spacelab one experiments on the behaviour of the vestibular system in space, by the so-called 'hop and drop' and 'drop and shock' methods. Scheduled for January 1986, it will also include NASA's Life Sciences Minilab. Twenty-five Life Sciences experiments have been chosen by NASA for Spacelab four. These will study the biomedical problems associated with human spaceflight and the effects of microgravity on living organisms. Research on the safety, wellbeing and work efficiency of people in space will be carried out in this spaceborne biological laboratory. The test subjects will include the crew members themselves. Three of the investigators are non-Americans, two coming from Europe and one from Australia.

The **International Microgravity Laboratory** (IML) is an international cooperative venture. It will be a Spacelab, fitted out with Life Sciences and Material Sciences equipment, provided by the participating countries or organisations such as ESA, NASA, Canada and Japan. In view of the considerable interest in microgravity research, NASA plans to provide free launches and payload integration in return for the use of any equipment aboard. For instance, ESA might provide the Biorack and/or Anthrorack. The latter is under development for eventual use in Spacelab for human physiological research. At least three missions are planned to form the IML programme, with periodic launches at intervals of about 18 months.

A **Japanese Spacelab** mission and **Spacelab ten** are still in the planning stages. The former will carry Material Sciences facilities developed in Japan and will be devoted to the microgravity sciences. Spacelab ten is envisaged as a NASA Life Sciences mission, probably reflying much of the equipment of Spacelab four. ESA has been invited to take part in it.

The **D-2** mission is a second German-managed Spacelab mission, concentrating on studies under microgravity conditions. This mission is at an early stage of development and the payload complement has not been decided. Possible major experiments would be the MSDR and the Sled. Microgravity investigations from other countries are also being considered. Much of the D-1 payload will be reflown. It

The Grille Spectrometer measures trace constituents in the Earth's atmosphere by analysing the Sun's infra-red radiation that passes through the atmosphere at each sunrise and sunset seen from the Orbiter (that is, by observing the Earth's atmospheric limb). The experiment first flew on Spacelab one and will be re-flown on EOM-1 under more favourable lighting conditions.

This artist's impression shows the Shuttle Infra-red Telescope Facility operating within the cargo bay.

is likely that the D-2 mission will take the place of the originally planned German D-4. The latter, using the igloo and four pallets, will probably now be a NASA-sponsored Astronomy mission, but the Federal Republic of Germany will retain a 50% interest in the payload through its GIRL telescope and associated experiments. GIRL (German Infra-Red Laboratory) consists of an infra-red telescope and up to four focal-plane detectors mounted in a cryostat containing superfluid helium. GIRL will use the IPS to provide the required pointing accuracy.

In addition to these Spacelab-dedicated missions, pallet elements of the Spacelab or pallet-like structures will fly on many 'mixed' missions. In these, the Orbiter will carry a mixed payload of, for example, communications satellites along with the pallet carrying experiments.

The present NASA philosophy is to concentrate on missions that are devoted to a particular scientific discipline, or to different disciplines with similar requirements. This is a simpler approach than that adopted on Spacelab one where the objective was to demonstrate Spacelab's usefulness in a variety of disciplines. Already developed instruments will be flown and reflown (as is already the case for EOM-1). This approach derives benefits from existing instruments in the most cost-effective way. This leads to the concept of 'discipline laboratories' that make use of proven hardware. Several such missions are under consideration.

Other discipline laboratories are planned as dedicated Spacelab missions. The Solar Optical Telescope (SOT) is a one-metre-diameter telescope which will use the Instrument Pointing System. It will be used to explore basic solar phenomena, such as plasma heating and energy redistribution, with measurements from the near infra-red to the far ultra-violet. It will probably be launched in 1989. Another one-metre telescope is the Shuttle Infra-Red Telescope Facility (SIRTF). It will be cryogenically cooled to temperatures as low as 10 Kelvin, with the detecting instruments at even lower temperatures (about 2 or 3 Kelvin). A first flight is contemplated for mid-1990; however it is quite likely that SIRTF will become a free-flying mission.

## Some planned discipline-oriented space laboratories

| Probable first launch date | Name | Purpose |
| --- | --- | --- |
| June 1984 | Large Format Camera (LFC) | To take synoptic, high-resolution images of the Earth's surface for cartography and mineral exploration |
| August 1984 | SPARTAN | To perform Astrophysics experiments with relatively simple equipment |
| October 1984 | Shuttle Radar Laboratory (SRL) | To acquire radar images of the Earth's surface |
| November 1984 | Material Sciences Laboratory (MSL) | To study Material Sciences phenomena in microgravity conditions |
| March 1986 | ASTRO | To obtain ultra-violet data on astronomical objects (including Halley's comet) by three independent telescopes using the IPS |
| July 1986 | SUNLAB | To study the Sun and the solar atmosphere, by reflying four solar instruments from the Spacelab two mission |
| March 1988 | Shuttle High Energy Astrophysics Laboratory (SHEAL) | To make high-accuracy measurements of X-radiation by four instruments |
| June 1988 | Space Plasma Laboratory | To conduct active experiments in the Earth's ionised atmosphere |

These discipline-oriented laboratories will fly repeatedly, with only moderate up-dating. An eventual step is to make some of them part of a Space Station, either as a dedicated manned laboratory or as an attached, unpressurised observatory.

For later Spacelab flights, Canada is planning three instruments. One observes airglow emitted by molecules and ions at different heights in the uppermost atmosphere. Another investigates the composition and energy distribution of magnetospheric ions with an extremely sensitive mass spectrometer. The third concerns electrostatic and electromagnetic waves launched into ionospheric plasma.

## EURECA – Europe's free-flying laboratory

As a follow-on to Spacelab, ESA has undertaken the development of EURECA, the European Retrievable Carrier. It is a free-flying satellite, launched and retrieved by the Space Shuttle. It has a mass of about 4000 kilograms at launch and can carry a payload of about 1000 kilograms. One kilowatt of continuous power is available for the experiments; active cooling is provided. Data can be transmitted to ESA's S-band ground stations at a rate of 256 kilobits per second. When ground stations cannot be 'seen', up to 128 megabits can be stored aboard EURECA.

In many ways EURECA exhibits the better qualities of manned and unmanned flight as embodied in Spacelab and conventional expendable satellites, respectively. The disturbing influence of human beings during the mission is avoided, while still retaining the possibility of manned intervention during launch. This is particularly relevant for Astronomy observations and for microgravity experiments where people are used to set up the experiment but where man-made perturbations during observation are undesirable. Also, the safety requirements associated with a fully manned system can be considerably relaxed with EURECA. Many experiments require longer flight durations than those provided by Spacelab, yet conventional satellites require a high level of redundancy of components to ensure a high reliability of operation. EURECA gives a 6- to 8-month operational period with the added advantage of retrieval, repair, refurbishment and re-use. Furthermore, the total launch costs are considerably less than for Spacelab since the system is relatively short (about 2.5 metres long) and a smaller payload mass is involved.

The operational scenario is as follows. The experiments are integrated onto the EURECA platform and their operation is checked. The whole package is then sent, by a conventional aircraft, to the Kennedy Space Center where it is fitted into the Orbiter. After launch into low Earth orbit, EURECA is removed from the cargo bay by means of the Remote Manipulator System (RMS) and released as a free-flying laboratory. A 400-newton thruster of its built-in Orbiter Control System (OCS) is then fired, taking EURECA to its operating altitude of about 500 kilometres. The solar arrays are deployed to provide power

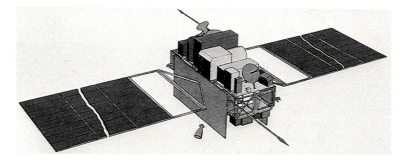

The EURECA spacecraft, on its first flight, will carry automatic Material Sciences and Life Sciences facilities powered by the 15 metre solar arrays. Heat is rejected from the large radiator shown in grey.

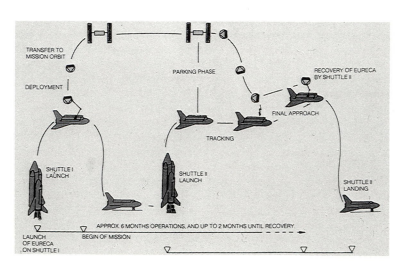

This flight profile illustrates the free-flying nature of EURECA. Launch and retrieval will be made during separate Shuttle flights that are 6- to 8-months apart.

for the system and the experiment phase begins. To minimise costs, EURECA's solar arrays are fixed so that the vehicle always faces the Sun; a pointing accuracy of about ± one degree will be maintained.

After about 6 months, EURECA is returned to low Earth orbit, by again using its OCS. The Orbiter makes a rendezvous with it and the RMS is used to retrieve EURECA which is then brought back to Earth for payload exchange and use in a later mission.

During its operational phase the microgravity environment is good; residual accelerations are less than $10^{-5}g$. EURECA is, therefore, ideal for long-duration microgravity experiments, such as the growth of certain crystals and the investigations of the growth histories of botanical specimens. In fact, the first mission of EURECA, scheduled for launch in October 1987 aboard the Orbiter Challenger, with retrieval in May 1988 by Columbia, is mainly devoted to microgravity research. To this end, several automatic facilities are being developed by ESA. Examples are furnaces and a system for exposing biological samples to the environment of space. Some thirty-five experiments will use these facilities. In addition, four science experiments taking advantage of the solar pointing and three technology experiments will be carried. One of these concerns a space link whereby data from EURECA are transferred to Earth at high rates via the large geostationary satellite L-SAT, now known as Olympus. Follow-on microgravity missions are planned and it is likely that they will involve international cooperative payloads. EURECA, complementing Spacelab, plays a key role in ESA's future microgravity research programme.

The EURECA vehicle is also very suitable for investigations in other disciplines. In this context experiments in Astronomy, Solar Physics, Atmospheric Physics, Earth Observations and Technology can be performed using EURECA. In such cases, the data transmission rate needs to be increased, the pointing accuracy improved and continuous viewing of a chosen target made possible. Techniques enabling EURECA to rendezvous and dock with another orbiting vehicle will be developed.

EURECA can be regarded as an extremely useful tool in its own right. Additionally, it represents a bridge between Spacelab and the Space Station. It is, in fact, a stepping stone to the future.

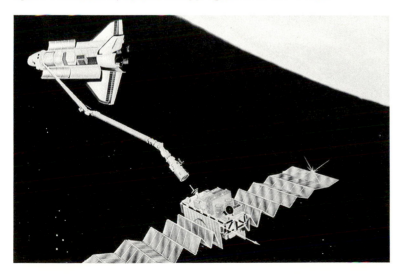

EURECA is being deployed from the Orbiter using the Remote Manipulator System. After release it is moved to a safe distance and its integral rocket motor ignited to place it in a 500 kilometre altitude orbit.

# Space Stations for the future

The Space Shuttle Program in general and the Spacelab Programme in particular form the foundations for other programmes, leading eventually to a permanently manned Space Station for science and industry. Several developments are foreseen by NASA and ESA, initially in low inclination orbits (28.5 degrees) with launches from Kennedy Space Center. By the end of the twentieth century there could be a large instrumented platform in a polar orbit using Vandenberg Air Force Base in California as the launch site. This orbit is extremely valuable for remote-sensing and space environment studies.

As a first step, the Space Shuttle has proved its ability to take astronauts to the orbit of an ailing satellite. This feature was demonstrated for the first time in April 1984. During the eleventh Shuttle flight an attitude control module was replaced on the Solar Maximum Mission Satellite and some instruments were repaired.

The capabilities of the Space Transportation System (STS) itself can be extended and its uses increased. Extra fuel cells would be needed to provide more electricity; extra propellants for the Orbital Maneuvering System (OMS) would have to be supplied. In this way, the mission could be extended up to about 60 days. However, a correspondingly reduced payload mass would result. Alternatively, more electrical power could be generated by an array of solar cells. These are photovoltaic cells which transform the Sun's radiant energy into electrical energy in the form of a DC voltage. Left in orbit, such an array could also provide power for an unmanned free-flying satellite, launched, visited and tended by the Space Shuttle.

It would be possible to contemplate a manned Spacelab module being taken out of the Orbiter's cargo bay to become a free-flying satellite itself. It would need an attached solar array for electrical power. Excess heat would have to be radiated from a panel. Platforms derived from Spacelab or from the EURECA-concept pallets could support additional experiments.

The next natural step would be to group several such modules and pallets together, forming an embryonic Space Station. Interconnected modules, some being used for living and some for working, could form a network with unmanned platforms carrying instruments. This could grow and develop into a base which is manned continuously and where the exploitation of space becomes a reality.

A permanently manned Space Station could be used for scientific research and commercial applications over a wide range of disciplines. The advantages of such a Space Station include a long time – up to 10 years – in the space environment, the capability for operating very large apparatus, and the large amount of electrical power available. The Space Station could be in low Earth orbit or, eventually, in geostationary orbit.

Large instruments would be invaluable for astronomical observations. Long-term investigations could be made of the Sun and the near-Earth environment. Observations of the Earth's surface and oceans would be extremely useful through the different seasons. Uninterrupted Life Sciences investigations would be possible. Technological uses would be the deployment of large structures in space and the development of energy conversion techniques. The new

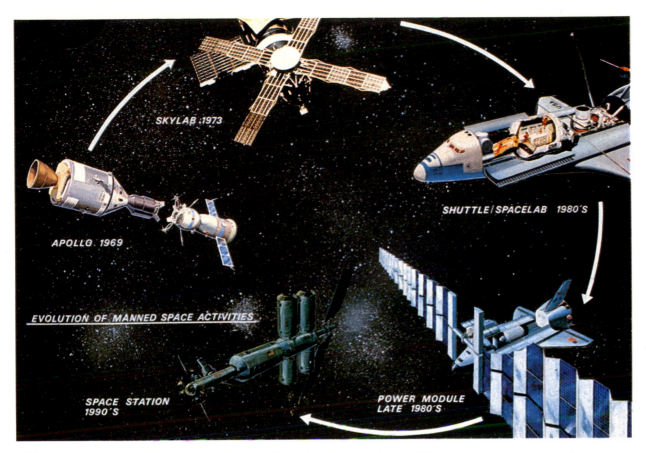

SKYLAB 1973

APOLLO 1969

SHUTTLE/SPACELAB 1980'S

EVOLUTION OF MANNED SPACE ACTIVITIES

SPACE STATION 1990'S

POWER MODULE LATE 1980'S

subjects of computer science and robotics will evolve to meet the needs of the Space Station. Yet it is the commercial applications that hold the greatest promise. The 'industrialisation of space' has long been a dream of planners. The Space Station could well see the realisation of those dreams through the processing of exotic materials and the preparation of pharmaceuticals or vaccines. Powered by huge arrays of solar cells, this could also be a base and a factory for constructing parts for other Space Stations. An example of such a Space Station would be a solar power station to transmit energy to the Earth as microwaves.

The Space Station will also act as a staging post for transferring payloads into other orbits, such as the commercially important geostationary orbit. For this, a new rocket – or 'Orbital Transfer Vehicle (OTV)' – would have to be developed. This could also be used for lunar and planetary missions that require enough energy to escape from the Earth's gravity. This OTV could be serviced and stored in the Space Station.

Early next century, a Space Station in an orbit at about 400 kilometres altitude could provide hotel accommodation for visitors reaching it aboard the Space Shuttle. Food would need to be grown in orbit; waste products could, perhaps, be transformed into fertilisers. Tourist attractions could be the observation of the Earth and the Universe, microgravity gymnastics and walks in space. The 'colonisation of space' could follow. Then, many people would make their homes in space, living and working in a man-made environment.

Apollo and Skylab in the past, the Space Shuttle and Spacelab at present, and Shuttle activities in the future illustrate the evolution of manned spaceflight to a permanently manned Space Station. Spacelab could initiate Europe's involvement in this exciting venture.

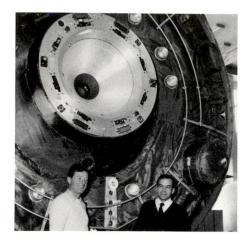

(Above) The two French 'spationautes', Jean-Loup Chrétien and Patrick Baudry, pose in front of a full-scale model of Salyut at the cosmonaut training centre at Star City, USSR. Chrétien spent 6 days aboard Salyut 7 performing medical experiments. His back-up for the Soviet flight, Baudry, will fly on a future Space Shuttle mission to operate a French experiment which monitors blood flow in the body during weightlessness.

(Below) The Soviet Space Station Salyut-6 has two docking ports for Soyuz crew-transfer vehicles, living quarters and a laboratory. Power is provided by the extended solar arrays. Up to five cosmonauts can be accommodated over many months.

# The USSR Space Station Programme

The Soviets have, during the past 15 years, shown to the world that a Space Station is an important part of their space programme. They have used Soyuz spacecraft to launch many astronauts and to rendezvous and dock with a Salyut Space Station. This is in an orbit of 52 degrees inclination at an altitude around 300 kilometres. The astronauts return to Earth in a small, re-entry capsule, finally using parachutes and rockets to land.

The Soyuz spacecraft, first flown with an astronaut in 1967, can carry one, two or three astronauts. It has a mass of about 6500 kilograms, is 7 metres long and 2.7 metres in diameter. The first Salyut was launched in April 1971. It is 13 metres long, its maximum diameter is 4.2 metres, and its mass is approximately 19 000 kilograms. Having two habitable modules, namely a working and living compartment and a transfer and docking compartment, this can house up to five astronauts for many months. Powered by solar arrays, the Salyut Space Station has equipment for several scientific, technological and medical studies, including furnaces for processing materials and a multispectral camera. It also has an unpressurised section for both propulsion and mounting instruments.

Not only Soviet cosmonauts, but also astronauts from several countries of Eastern Europe and from Mongolia, Vietnam and Cuba, have flown in space on Soyuz and Salyut. A French 'spationaute' – Jean-Loup Chrétien – has also flown aboard Salyut. Astronauts who have spent as much as 6 months in orbit have reported that reacclimatising to the one-$g$ conditions on the Earth's surface is both unpleasant and difficult.

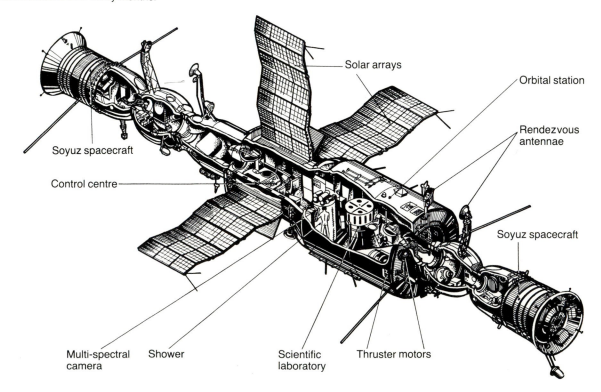

Solar arrays

Orbital station

Rendezvous antennae

Soyuz spacecraft

Soyuz spacecraft

Control centre

Multi-spectral camera

Shower

Scientific laboratory

Thruster motors

In June 1983, a new, large supply spacecraft was launched to dock with the Salyut-7 Space Station. With a length of 13 metres and a total mass of 20 000 kilograms, this contains almost three times as much cargo space as the 2300 kilograms available on the earlier re-supply vehicle, Progress. It has solar cells over an area of 40 square metres. Furthermore, it can modify the orbit of the Space Station. Also it contains a capsule in which up to 500 kilograms of materials produced in space, equipment to be re-used, or film can be returned to Earth.

A new Space Station is planned for launch in 1986. A central core capsule, similar in size to the present Salyut, will provide the living volume. Six other spacecraft will be attached to this. Two will be laboratories for research in Astronomy, Earth Observations, Material Sciences and Life Sciences; two will be Soyuz spacecraft which will have brought two or three astronauts to the Space Station; and two will be unmanned supply spacecraft of the Progress type. The astronauts will eventually return to Earth in a landing capsule which forms a small part of the Soyuz spacecraft. For this programme, a new, large rocket is being developed which will be even more powerful than the US Saturn-V rocket.

Even so, the concept of re-usable space hardware is being actively explored. To this end, the Soviets launched a small (1000-kilogram) re-usable winged space vehicle in June 1982 and March 1983. Somewhat resembling a small (4.6-metre-long) Space Shuttle, this orbited the Earth once. It safely returned to Earth by deploying a parachute and landing in the Indian Ocean. There it was recovered by a ship of the Soviet Navy. The third flight was in December 1983, with splashdown in the Black Sea. This programme could be a prelude to a Soviet Space Shuttle. It is thought that such a Shuttle would be rather larger than the US Shuttle, being 65 metres tall on the launch pad (as against 56 metres for the US Space Shuttle). The two boosters are believed to use liquid fuel and the external tank to contain cryogenic fuel. The three main rocket engines are an integral part of this tank. Thus neither the tank nor the engines would be re-usable. This Space Shuttle 'look-alike' could be launched from Tyuratam, near the Aral Sea, and make its return landing on a runway there. Future developments are eagerly awaited.

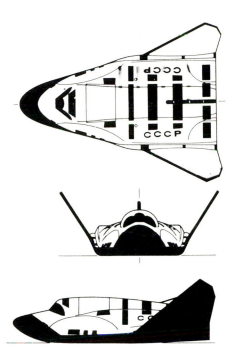

Three diagrams of the small Soviet re-usable winged vehicle that has flown in space. It could herald a Soviet version of the Space Shuttle.

## The US Space Station

President Ronald Reagan announced in his 1984 State of the Union address that the USA would have a permanently manned Space Station within a decade. Such a permanent presence in space has always been considered to be a significant milestone in space exploration. The US Space Station will be designed and built, under the responsibility of NASA, at an initial cost of approximately US$8 billion. The modular elements of the Space Station will be launched into low Earth orbit by the Space Shuttle and assembled there. Initially the US Space Station will support a crew of six to eight persons to be changed every few months, again using the Space Shuttle. It will also provide work space and the necessary services (power, thermal control and data handling) for both scientific and applied tasks in orbit. Like all advanced developments the uses of the Space Station will evolve as its capabilities grow and as people become more experienced in handling it.

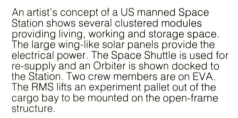

An artist's concept of a US manned Space Station shows several clustered modules providing living, working and storage space. The large wing-like solar panels provide the electrical power. The Space Shuttle is used for re-supply and an Orbiter is shown docked to the Station. Two crew members are on EVA. The RMS lifts an experiment pallet out of the cargo bay to be mounted on the open-frame structure.

The US Space Station concept is essentially a manned base in a low-inclination orbit (28.5 degrees) at an altitude of about 450 kilometres. It will provide both living quarters for the crew and laboratory space for accommodating experiments; it will have provisions for attitude control, data processing and electrical power. People in the Space Station will be able to look after free-flying satellites or platforms alongside the Space Station. They could also visit them from time to time using a Manned Maneuvering Unit (MMU) similar to that already used when repairing the Solar Maximum Mission Satellite. In this way the equipment could be updated or changed, but the actual scientific observations could be made from the instrumented satellites without any perturbations being introduced by Man.

As part of the proposed US Space Station Program, President Reagan suggested that 'America's friends and allies' could participate in the NASA programme. The cooperation could range from the development of parts of the Space Station itself to its use once completed. Obvious choices for such participation are Europe, Canada and Japan; in Europe this could be as individual countries or as a concerted effort through ESA.

## Some possibilities for Europe

In February and March 1984, NASA's Administrator, James Beggs, visited the capitals of possible participating countries to explain the forms which such cooperation might take. The cooperation of international partners with NASA would involve a commitment to the three phases of the programme, namely, study, development and operation. Three possible schemes are apparent. The first involves an approach in which all partners work in a programme that is a joint venture. In the second form of cooperation, partners contribute other parts which will be added onto the basic Space Station. Use by that partner's experimenters of the entire Space Station should then be

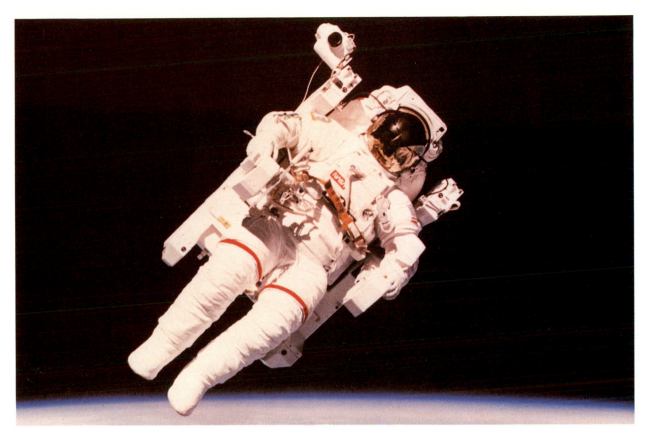

possible on preferential terms. The third possibility is that the developer of a certain part for the Space Station would retain its exclusive use. In this case, the partner would not have preferential treatment in obtaining access to NASA's facilities. In all cases, some 'bartering' can be envisaged, either for hardware or for services rendered, to ensure that no financial transactions are needed that involve foreign exchange.

Whatever cooperation scheme is adopted, NASA would consider the contribution to be over and above the investment of US$8 billion. The added contribution would be seen as an improvement to the Space Station. In this way, NASA can proceed with its plans without being too concerned about the amount of foreign involvement.

It is reasonable to assume that any European contribution to the US Space Station Program will draw heavily on the experience gained from Spacelab and EURECA. In fact, the modular elements of Spacelab and EURECA are the ideal building blocks for future Space Station hardware.

Spacelab is a habitable module. By extending its capability for staying in space to that required for a Space Station, it becomes an obvious candidate for one of the additional modules envisaged by NASA. It could act as the basis for a crew accommodation module, a storage facility, or a laboratory for conducting experiments operated or tended by men. Knowing the strong European interest in the discipline of Material Sciences, Spacelab is a worthy candidate for a microgravity laboratory docked permanently to the basic Space Station.

Astronauts on-board a Space Station will need to visit various parts of the Station. This will be done using a personal propulsion device such as the Manned Maneuvering Unit shown. Here, Bruce McCandless rides his nitrogen-propelled, hand-controlled aid to EVA near the Orbiter Challenger.

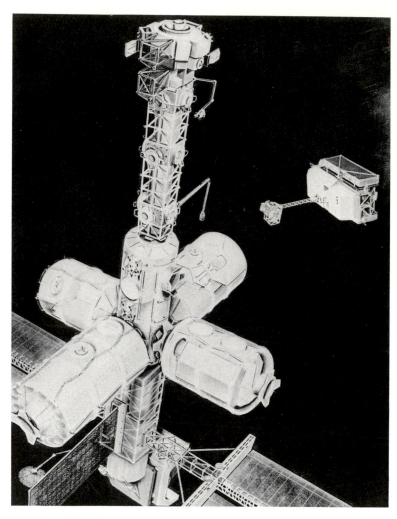

This European concept of a Space Station shows how Spacelab modules might be clustered to form its main core. An extra Spacelab cylindrical section can increase the module size. The long mast structure allows the docking of experiment carriers such as the EURECA-like free-flyer to the right. A Teleoperator Maneuvering System (TMS) used for moving and servicing equipment is shown docked to one end of the mast.

A EURECA-based system is a natural candidate for either an unmanned platform or a platform attached to the Space Station. In the former application, one or more EURECA structures could be used as the basis for an instrumented, free-flying platform in space. Large solar arrays could be used to supply power in the kilowatt range for the free-flyer. It could be returned to Earth (as in the present concept) for refurbishment or be maintained in low Earth orbit. Experiments could be changed in a similar way.

Thus, Europe possesses the capability and experience for contributing two of the key elements of a Space Station. Added to this is the known European interest and experience in robotics, so that a vehicle for servicing equipment, either manned or unmanned, becomes another candidate for a European contribution. Since transport to the Space Station will be by the Space Shuttle, for crews, hardware and supplies, the compatibility of both existing European systems with the US Space Transportation System is an advantage.

Much work is being carried out, by ESA and within its Member States, to investigate the many possibilities open to Europe. A joint project between the Federal Republic of Germany and Italy, project Columbus, is particularly noteworthy. Based on developments of

(Left) Two possible unmanned ESA contributions to Space Station activities are depicted in this sketch. They are a remotely controlled service vehicle, with robotic arms and a large free-flying platform based on Spacelab pallets or EURECA.

(Below) The Columbus project includes a Service Vehicle which may be either manned or unmanned (shown on the right), a Resource Module with its power-producing solar arrays (shown on the left), a pressurised Research Module and an unmanned Platform. The latter two items which carry experiments, may be docked to the US Space Station or may be free-flying. All elements are based on Spacelab or EURECA designs.

Spacelab, EURECA and SPAS (Shuttle Pallet Satellite), it would use the Space Shuttle for launch and resupply. The main elements are a pressurised microgravity research laboratory, a space platform for carrying experiments, a 'resource module' for providing the free-flying laboratory and platform with all its needs, and a servicing vehicle. All these elements could become part of the US permanent Space Station.

Another possibility is that the Europeans might decide that they 'want to go it alone'. In this case, parts of Spacelab and EURECA could be used as the basic station or platform. However, such an independent approach requires a European launch capability. An advanced version of the European rocket launcher, Ariane-V, could satisfy this requirement. The mini-Shuttle, Hermes, proposed by France, might then be used to take astronauts to and from the European Space Station. Hermes has a mass of about 15 000 kilograms and a payload capability of some 7000 kilograms. The manned version could have a crew of four and a cargo of some 4000 kilograms. It is 16.8 metres long, 5.5 metres high and has a wingspan of 10.4 metres. Launched by Ariane-V, Hermes would re-enter and land like the American Orbiter.

Europe must decide how to cooperate with the USA to make the NASA Space Station a truly international venture. The path chosen will determine the way forward to the end of the twentieth century and beyond.

The Hermes mini-Shuttle could be used for ferrying men and supplies between Earth and a European Space Station. It would be launched by an Ariane-V rocket and land like a glider.

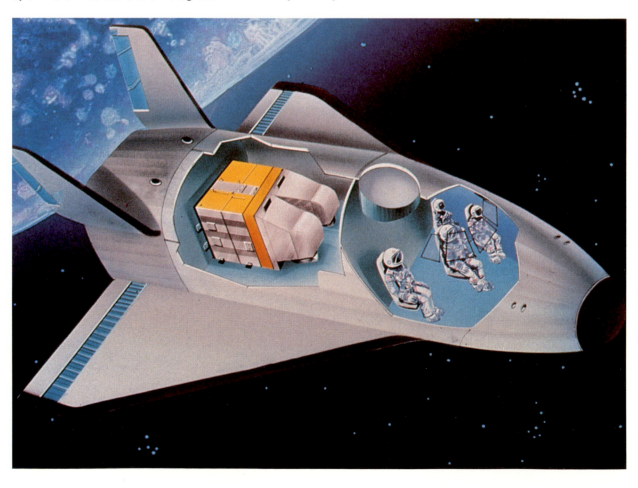

# VIEWS OF THE SPACELAB ONE PAYLOAD CREW

The Spacelab one scientist–astronaut team is shown here beside a model of the Orbiter with Spacelab in its cargo bay. From left to right they are: *Robert Parker* (US Mission Specialist), *Owen Garriott* (US Mission Specialist), *Ulf Merbold* (European Payload Specialist), *Michael Lampton* (back-up US Payload Specialist), *Wubbo Ockels* (back-up European Payload Specialist) and *Byron Lichtenberg* (US Payload Specialist).

## Robert Parker

My involvement with space laboratories began with my Earth-bound participation in Skylab. Even before the end of Skylab, this involvement grew to include discussions with strangers (soon to be long-term associates) from ESRO who had similar interests in and plans for using space. As the hardware design matured, many of us became involved with the question of how to operate Spacelab and our attention was periodically drawn to Payload Operations Working Groups, combining contractors, ESA and NASA. In the meantime, NASA and ESA were studying the question of how to do science in this new vehicle. Reviewing our experience in Skylab and anticipating the arrival of Payload Specialists we proceeded to explore new ways of interacting, training, and operating instruments. Attempting to put into practice the lessons learned, I next joined as a Mission Specialist with the group of US and European Payload Specialists to form the Payload Crew for Spacelab one. Training began in the laboratories of the many Principal Investigators and yet another group of associates was added to the cast. As training progressed from the science of the individual instruments towards the integrated operation of the instruments and the vehicle itself, more and more operations engineers from ESA/SPICE and NASA joined our diverse and growing team.

With the vehicle and the experiment hardware arriving at the Kennedy Space Center, both groups of participants came together and, with teams of NASA and ESA contractors, pushed the total payload through a surprisingly smooth integration and check-out. At Huntsville and Houston, simulations put the final pieces into place. On 28 November 1983, six of us lifted off. With the continuing support of the rest of the team, in control rooms scattered across two continents, we saw to it that our Spacelab and her experiments did their job and did it well.

For most of us the efforts involved in this saga comprised a significant fraction of our professional lives. During that time our personal lives were also changed in many ways, not the least of which was our new friendships with so many of the team members. I suspect that for some time to come we will look back with keen satisfaction at a job well done, a team well respected and friends well prized.

# Owen Garriott

Spacelab has proven to be an excellent laboratory for conducting multi-disciplinary space science. Not only are special research facilities available, but the basic requirements of a manned environment are provided. Atmospheric pressure, composition, temperature and humidity are regulated, electrical power from the Orbiter is distributed, heat is transported back to the Orbiter cooling loops and communications and data management are provided. Most of these systems are controlled automatically, requiring minimal manual intervention by the crew.

Most of our time was spent performing the more appropriate tasks associated with payload operations. It was somewhat frustrating to know that only the surface could be scratched in reaching the full potential of our many experiments. Later flights and eventually the Space Station should provide the long-duration opportunities needed by so many of the experiments.

Substantial changes in the way manned space science is conducted have been made in the 10 years between Skylab and Spacelab. First, the computer revolution has permitted the automatic selection of filters, exposure times, many sequential operations and a whole host of other functions, while the human participant selects operational modes and interprets the data. Secondly, data exchange and communications between the flight crew and ground investigators have been greatly expanded. This has led to the third major change: the ground investigator is now the guiding figure in the real-time conduct of his experiment. In some cases experiments are fully controlled by ground command, whereas in other cases the flight crew and ground investigators work together as a team to achieve the objectives.

It has been a great pleasure to work with science investigators and crew members from all over the world. Spacelab has demonstrated not only a sparkling performance but also the viability of a major international cooperative effort. We hope that our work will be only the beginning.

# Ulf Merbold

The preparation for the Spacelab one mission and its fulfilment have truly been the greatest experience of my professional life. First, the mission was a great intellectual challenge. The experiments, from seven different scientific disciplines, provided a colourful mix that gave members of the Payload Crew the unique opportunity to broaden their scientific horizons. The multi-disciplinary flavour gave me the chance to introduce myself to new fields like Astronomy, Atmospheric Physics, Earth Observations, Biology and Medicine, and Plasma Physics. It was during the simulations that the scientists learned from us, because of our knowledge of Spacelab. The intense interaction between scientists in the POCC and the Payload Crew created a splendid team of hard-working and dedicated people. NASA allowed the scientists to communicate directly with a flight crew for the first time. As a result, all the problems encountered during the flight could be solved.

Secondly, the mission was a rewarding social experience. The international mix of Principal Investigators and their associates,

Payload Crew, and NASA and ESA engineers forced all of us to correct quite a few prejudices. The professional interactions created very close human relationships. This friendship was amplified during the flight, when six men spent ten days in orbit, confined in a spacecraft, without any moments of tension or friction.

Thirdly, the first Spacelab flight was a once-in-a-lifetime experience regarding personal emotions and beauty. We were all lucky to orbit mother Earth at an orbital inclination of 57 degrees. This allowed us to observe almost all populated areas of our beautiful planet. Within one hour we might see tropical thunderstorms, auroral displays extended over thousands of kilometres and the many different colours of the deserts. Most spectacular of all were the high mountains. What we did not see were borders between countries. Because it took us typically only a minute to fly from one European country to the next, I realised that we live on a small globe. The time of the mission was December. Never before did I feel so clearly the meaning of the Christmas message: 'Et in terra pax – Peace be on Earth'.

Last, but not least, it was surely a very special satisfaction for all of us Europeans that Europe's contribution to the new NASA Space Transportation System – Spacelab, furnished by the European Space Agency – worked flawlessly during the entire flight.

# Byron Lichtenberg

I should like to begin by expressing my deepest thanks to all the people who have worked on the Spacelab one mission for so many years. It has truly been one of the most wonderful experiences of my life. The very close personal and working relationships that I have established with many friends in ESA and NASA have broadened my horizons and enriched my life. I believe and hope that the outcome of Spacelab one has demonstrated the very great capabilities that Spacelab and the Space   Shuttle have for doing top-quality research in space.

Making the Principal Investigators primarily responsible for the scientific training of the Payload Crew worked extremely well. The chance to work with such eminent scientists was a once-in-a-lifetime opportunity. Also, working side by side with their laboratory technicians and co-investigators gave us a clear insight into the important areas of their experiments. We planned, for the first time on a spaceflight, routine direct communications between the Principal Investigators and the science crew on-board.

Spacelab gave a wonderful environment in which to conduct science. The life support, electrical power, communications and data systems of Spacelab worked as advertised. For the first flight of a system, very few problems were encountered. We were able to devote almost all our time to conducting science operations, which was terrific. We learned that it was important to have some extra time in the schedule to allow for adaptation to weightlessness and to solve problems. We discovered that the concept of direct communication between the Principal Investigator and Payload Crew is very beneficial.

The experience of being weightless and doing science 240 kilometres above the Earth's surface was remarkable. The only thing that I would change would be to protect better some of my off-duty hours so that I could use the opportunity to view even more the splendid Earth on which we live. The views from space made me realise just how small, fragile and beautiful our planet really is. If there is one personal insight that I gained from flying in space, it is that all of us on Earth must be responsible for our planet and take good care of it.

## Wubbo Ockels

Thinking back, I recall the very intense and rewarding team work involving the crew, scientists and engineers in the POCC and support teams during the ten days and 166 orbits of the first Spacelab mission. The communication between myself, as back-up Payload Specialist on the ground, and Ulf Merbold on-board allowed me to be mentally with him inside Spacelab.

The Payload Crew started training for Spacelab one in mid-1978. The method of implementation, the huge amounts of documentation, the difficulty of obtaining information, the number of people who needed to understand the mission plans, the concern about the crew's health and safety, and the different opinions on what the crew should do impressed me. It was all very different from a laboratory on Earth. Because the launch was postponed, I had the opportunity to train in the US as a Mission Specialist, for 1½ years. The Spacelab simulator at the Marshall Space Flight Center became an essential part of the preparations for the Spacelab one mission. It was during the many simulations that all members of the team developed an understanding of how they had to work together during a mission.

When the real flight started, on 28 November 1983, many people in the POCC felt as if it was another simulation. For the scientists, however, the difference between reality and simulation is absolute and having real data after all the years of preparation produced a euphoric mood. During the flight, all the capabilities available were used to make the mission a success. Many repairs were performed by the crew in orbit and several experiments were optimised.

We are now preparing for the German D-1 mission, another step in the direction of establishing a permanent laboratory in space in which scientists will work, be creative and make their inventions.

## Michael Lampton

Now that the excitement of the first mission is over, the numerous debriefings transcribed, the voluminous digital tapes processed and the public relations tours completed, what have we accomplished?

First, Spacelab one has demonstrated an extraordinary degree of cooperation. By this I mean that, in a multidisciplinary mission, the speed with which a unified plan can emerge from the often conflicting interests of the participating scientists is astonishing.

Secondly, we have shown that a large degree of operational flexibility is a firm requirement for productivity on-orbit, just as it is in a ground-based facility.

Thirdly, the mission established the usefulness of having the Principal Investigators working in close contact with their crew. Many of the experiments benefitted from this arrangement. We may take some satisfaction from having established a trend for all subsequent space missions in which close interactions in the processes of execution, observation and interpretation phases of research can only be beneficial.

# APPENDIX

This Appendix lists the experiments performed on the first Spacelab flight and indicates the degree of success achieved. It also provides the names of the Principal Investigator, his or her organisation and the experiment's mass. The location of each experiment within the Spacelab module (M) or on the pallet (P) is also indicated.

The coding of the experiments' numbers is explained as follows:

1 indicates the first Spacelab flight
N indicates choice by NASA
E indicates choice by ESA
S indicates a science experiment
A indicates an applications experiment
T indicates a technology experiment
three numbers specify the experiment

Success achieved during the first Spacelab mission is indicated as follows:

A = 100% successful
B = at least 75% successful
C = 25 – 50% successful
0 = 0% successful
(There were no experiments with success rates between 50 and 75%.)

| Title | First Spacelab flight experiment number | Principal Investigator and Institute | Location | Mass (kilo-grams) | Success achieved |
|---|---|---|---|---|---|
| **NASA experiments** | | | | | |
| *Astronomy and Solar Physics* | | | | | |
| Far Ultra-violet Space Telescope (FAUST) | 1NS 005 | C.S. Bowyer, University of California at Berkeley, USA | P | 89.6 | C |
| Active Cavity Radiometer (Solar Irradiance Monitor) | 1NS 008 | R.C. Wilson, Jet Propulsion Laboratory, USA | P | 21.6 | A |
| *Space Plasma Physics* | | | | | |
| Space Experiments with Particle Accelerators (SEPAC) | 1NS 002 | T. Obayashi, Institute of Space & Astronautical Sciences, Tokyo, Japan | P | 398.1 | B |
| Atmospheric Emission Photometric Imaging (AEPI) | 1NS 003 | S.B. Mende, Lockheed Palo Alto Research Laboratory, USA | P | 221.0 | B |
| *Atmospheric Physics and Earth Observations* | | | | | |
| Imaging Spectrometric Observatory (ISO) | 1NS 001 | M.R. Torr, Utah State University, USA | P | 246.3 | B |

*Life Sciences*

| | | | | | |
|---|---|---|---|---|---|
| Radiation environment mapping | 1NS 006 | E.V. Benton, University of San Francisco, USA | M | 2.6 | A |
| Preliminary characterisation of persisting circadian rhythms during spaceflight: *Neurospora* as a model system | 1NS 007 | F.M. Sulzman, State University of New York at Binghamton, USA | M | 3.4 | A |
| Minilab comprising: | 1NS 100 | | M | 354.3 | |
| Nutation of *Helianthus annuus* in a microgravity environment | 1NS 101 | A.H. Brown, University of Pennsylvania, USA | | | B |
| Vestibular experiments | 1NS 102 | L.R. Young, Massachusetts Institute of Technology, USA | | | A |
| The influence of spaceflight on erythrocyte kinetics in Man | 1NS 103 | C. Leach, NASA Johnson Space Center, USA | | | A |
| Vestibulo-spinal reflex mechanisms | 1NS 104 | M.F. Reschke, NASA Johnson Space Center, USA | | | A |
| Effects of prolonged weightlessness on the humoral immune response in Man | 1NS 105 | E.W. Voss, Jr., University of Illinois, USA | | | A |

*Technology*

| | | | | | |
|---|---|---|---|---|---|
| Tribological experiments in zero-gravity | 1NT 011 | C.H.T. Pan, Columbia University, USA | M | 72.3 | A |

**ESA experiments**

*Astronomy and Solar Physics*

| | | | | | |
|---|---|---|---|---|---|
| Very Wide Field Camera | 1ES 022 | G. Courtès, Laboratoire d'Astronomie Spatiale, Marseille, France | M | 99.8 | B |
| Spectroscopy in X-ray astronomy | 1ES 023 | R.D. Andresen, ESA (ESTEC), Netherlands | P | 20.5 | A |
| Solar spectrum from 170 to 3200 nanometres | 1ES 016 | G. Thuillier, Service d'Aéronomie du CNRS, Verrières-le-Buisson, France | P | 31.8 | A |
| Measurement of the solar constant | 1ES 021 | D. Crommelynck, Institut Royal Météorologique de Belgique, Belgium | P | 6.6 | A |

*Space Plasma Physics*

| | | | | | |
|---|---|---|---|---|---|
| Phenomena Induced by Charged Particle Beams (PICPAB) | 1ES 020 | C. Beghin, CNRS, Orléans, France | M/P | 43.8 | B |

| | | | | | |
|---|---|---|---|---|---|
| Low-energy electron flux and its reaction to active experimentation on Spacelab | 1ES 019A | K. Wilhelm, Max Planck Institut für Aeronomie, Hanover, Federal Republic of Germany | P | } 26.1 | A |
| DC magnetic field vector measurement | 1ES 019B | R. Schmidt, Space Research Institute of the Austrian Academy of Sciences, Graz, Austria | P | | A |
| Isotopic stack – measurement of heavy cosmic ray isotopes | 1ES 024 | R. Beaujean, Institut für Reine und Angewandte Kernphysik der Universität Kiel, Federal Republic of Germany | P | 22.2 | A |
| *Atmospheric Physics and Earth Observations* | | | | | |
| Grille Spectrometer | 1ES 013 | M. Ackerman, Institut d'Aéronomie Spatiale de Belgique, Belgium | P | 137.4 | C |
| Waves in the hydroxyl emissive layer | 1ES 014 | M. Hersé, Service d'Aéronomie du CNRS, Verrières-le-Buisson, France | P | 11.5 | A |
| Investigation of atmospheric hydrogen and deuterium through measurement of their Lyman-alpha emission | 1ES 017 | J.L. Bertaux, Service d'Aéronomie du CNRS, Verrières-le-Buisson, France | P | 13.2 | B |
| Metric Camera experiment | 1EA 033 | M. Reynolds, ESA, Toulouse, France | M | 155.4 | B |
| Microwave Remote-Sensing Experiment (MRSE) | 1EA 034 | G. Dieterle, ESA, Toulouse, France | P | 166.3 | 0 (active) A (passive) |
| *Life Sciences* | | | | | |
| Advanced Biostack experiment | 1ES 027 | H. Bücker, Institut für Flugmedizin/ Abteilung für Biophysik, Frankfurt/Main, Federal Republic of Germany | M/P | 7.8 | A |
| Micro-organisms and biomolecules in hard space environment | 1ES 029 | G. Horneck, Institut für Flugmedizin/ Abteilung für Biophysik, Frankfurt/Main, Federal Republic of Germany | P | 4.6 | A |
| Effects of rectilinear accelerations, optokinetic and caloric stimulations in space | 1ES 201 | R. von Baumgarten, Johannes Gutenberg Universität, Mainz, Federal Republic of Germany | M | 47.9 | B |

| | | | | | |
|---|---|---|---|---|---|
| Measurement of central venous pressure and determination of hormones in blood serum during weightlessness | 1ES 026 and 1ES 032 | K. Kirsch, Physiologisches Institut der Freien Universität, Berlin, Federal Republic of Germany | M | 3.7 | A |
| Mass discrimination during weightlessness | 1ES 025 | H. Ross, University of Stirling, United Kingdom | M | 4.5 | A |
| Three-dimensional ballistocardiography in weightlessness | 1ES 028 | A. Scano, University of Rome, Italy | M | 2.6 | A |
| Personal miniature electrophysiological tape recorder | 1ES 030 | H. Green, Clinical Research Centre, Harrow, United Kingdom | M | 5.8 | A |
| Effect of weightlessness on lymphocyte proliferation | 1ES 031 | A. Cogoli, Eidgenössische Technische Hochschule, Zurich, Switzerland | M | 5.5 | A |
| *Material Sciences* Material Sciences Double Rack (MSDR) comprising: | 1ES 300 | | M | 534.7 | |
| Mirror Heating Furnace Zone crystallisation of silicon | 1ES 321 | R. Nitsche, Kristallographisches Institut der Universität Freiburg, Federal Republic of Germany | | | A |
| Growth of CdTe by travelling heater method | 1ES 322 | R. Nitsche, Kristallographisches Institut der Universität Freiburg, Federal Republic of Germany | | | A |
| Growth of GaSb by travelling heater method | 1ES 323 | K.W. Benz, Universität Stuttgart, Federal Republic of Germany | | | A |
| Crystallisation of silicon spheres | 1ES 324 | H. Kölker, Wacker-Chemie, Munich, Federal Republic of Germany | | | A |
| Isothermal Heating Furnace Solidification of immiscible alloys | 1ES 301 | H. Ahlborn, Universität Hamburg, Federal Republic of Germany | | | A |
| Solidification of technical alloys | 1ES 302 | D. Poetschke, F. Krupp GmbH, Essen, Federal Republic of Germany | | | 0 |
| Skin technology | 1ES 303 | H. Sprenger, Maschinenfabrik Augsburg– Nürnburg AG, Munich, Federal Republic of Germany | | | 0 |

| | | | |
|---|---|---|---|
| Vacuum brazing | 1ES 304 | W. Schönherr, Bundesanstalt für Materialprüfung, Berlin, Federal Republic of Germany | A |
| | 1ES 305 | R. Stickler, University of Vienna, Austria | A |
| Solidification of monotectic alloys | 1ES 306 | H. Ahlborn, Universität Hamburg, Federal Republic of Germany | A |
| Reaction kinetics in glass melts | 1ES 307 | H.G. Frischat, Technische Hochschule, Clausthal, Federal Republic of Germany | C |
| Metallic emulsions Al–Pb | 1ES 309 | P.D. Caton, Fulmer Research Institute, Slough, United Kingdom | C |
| Bubble-reinforced materials | 1ES 311 | P. Gondi, Instituto de Fisica della Universitá, Bologna, Italy | A |
| Nucleation behaviour of eutectic alloys | 1ES 312 | Y. Malméjac, Centre d'Energie Atomique, Centre d'Etudes Nucléaires, Grenoble, France | C |
| Solidification of near-monotectic Zn–Pb alloys | 1ES 313 | H. Fischmeister, Montanuniversität Leoben, Austria | A |
| Dendrite growth and microsegregation | 1ES 314 | H. Fredriksson, Kungl-Tekniska Högskolan, Stockholm, Sweden | A |
| Composites with short fibres and particles | 1ES 315 | A. Deruyttere, Université Catholique de Leuven, Belgium | B |
| Unidirectional solidification of cast iron | 1ES 325 | T. Luyendijk, Laboratorium voor Metaalkunde, Delft, Netherlands | A |
| Low-temperature Gradient Heating Furnace<br>Unidirectional solidification of A1–Zn vapour emulsions | 1ES 316 | C. Potard, Centre d'Energie Atomique, Centre d'Etudes Nucléaires, Grenoble, France | A |
| Unidirectional solidification of eutectic alloys | 1ES 317 | J.J. Favier, Centre d'Energie Atomique, Centre d'Etudes Nucléaires, Grenoble, France | A |

| | | | |
|---|---|---|---|
| Growth of PbTe | 1ES 318 | H. Rodot,<br>CNRS Laboratoire<br>d'Aérothermique, Meudon,<br>France | A |
| Unidirectional<br>solidification of eutectics<br>(InSb-NiSb) | 1ES 319 | K.L. Müller,<br>Universität Erlangen,<br>Federal Republic of Germany | A |
| Thermodiffusion in<br>tin alloys | 1ES 320 | Y. Malméjac,<br>Centre d'Energie Atomique,<br>Centre d'Etudes Nucléaires,<br>Grenoble, France | A |

Fluid Physics Module

| | | | |
|---|---|---|---|
| Oscillation damping<br>of a liquid in natural<br>levitation | 1ES 326 | H. Rodot,<br>CNRS Laboratoire<br>d'Aérothermique, Meudon,<br>France | A |
| Kinetics of spreading of<br>liquids on solids | 1ES 327 | J.M. Haynes,<br>University of Bristol,<br>United Kingdom | A |
| Free convection in low<br>gravity | 1ES 328 | L.G. Napolitano,<br>Universitá degli Studi,<br>Naples, Italy | A |
| Capillary surfaces in low<br>gravity | 1ES 329 | J.F. Padday,<br>Kodak Limited, Harrow,<br>United Kingdom | A |
| Coupled motion of liquid–<br>solid systems in near-zero<br>gravity | 1ES 330 | J.P.B. Vreeburg,<br>National Aerospace<br>Laboratory, Amsterdam,<br>Netherlands | A |
| Floating zone stability in<br>zero gravity | 1ES 331 | I. Da Riva, Ciudad<br>Universitaria, Madrid, Spain | A |
| Interfacial instability and<br>capillary hysteresis | 1ES 339 | J.M. Haynes,<br>University of Bristol,<br>United Kingdom | A |

Single Experiments Using
Special Equipment

| | | | |
|---|---|---|---|
| Crystal growth of proteins | 1ES 334 | W. Littke,<br>Chemisches Laboratorium der<br>Universität Freiburg,<br>Federal Republic of Germany | A |
| Self-diffusion and<br>interdiffusion in liquid<br>metals | 1ES 335 | K.M. Kraatz,<br>Technische Universität,<br>Berlin, Federal Republic of<br>Germany | A |

| | | | | | |
|---|---|---|---|---|---|
| Adhesion of metals (UHV Chamber) | 1ES 340 | G. Ghersini, Centro Informazioni Studi Esperienze, Milan, Italy | | | A |

Other Material Sciences Experiments

| | | | | | |
|---|---|---|---|---|---|
| Crystal growth from solution – organic crystal growth | 1ES 332 | K.F. Nielsen, Technical University of Denmark, Lyngby, Denmark | M | } 25.3 | A |
| Crystal growth by coprecipitation in liquid phase | 1ES 333 | A. Authier, Laboratoire de Minéralogie – Cristallographie, Paris, France | M | | A |
| Crystal growth of mercury iodide by physical vapour transport | 1ES 338 | R. Cadoret, Laboratoire de Crystallographie et Physique de Matériaux, Clermont-Ferrand, France | M | 13.7 | A |

## Guaranteed reflights of experiments from Spacelab one

| Titles | Experiment number | Reflight |
|---|---|---|
| Very Wide Field Camera | 1ES 022 | Spacelab three, planned for January 1985 |
| Far Ultra-violet Space Telescope (FAUST) | 1NS 005 | |
| Space Experiments with Particle Accelerators (SEPAC) | 1NS 002 | Earth Observation Mission (EOM-1), planned for November 1985 |
| Atmospheric Emission Photometric Imaging (AEPI) | 1NS 003 | |
| Grille Spectrometer | 1ES 013 | |
| Waves in the hydroxyl layer | 1ES 014 | |
| Metric Camera experiment | 1EA 033 | |

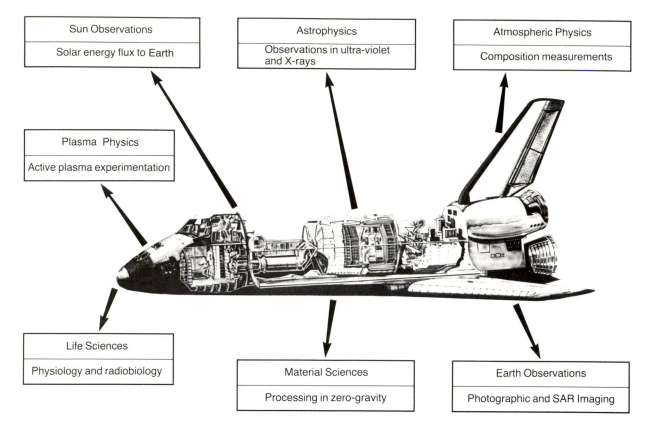

| Sun Observations | Astrophysics | Atmospheric Physics |
|---|---|---|
| Solar energy flux to Earth | Observations in ultra-violet and X-rays | Composition measurements |

| Plasma Physics |
|---|
| Active plasma experimentation |

| Life Sciences | Material Sciences | Earth Observations |
|---|---|---|
| Physiology and radiobiology | Processing in zero-gravity | Photographic and SAR Imaging |

Spacelab one pallet experiments

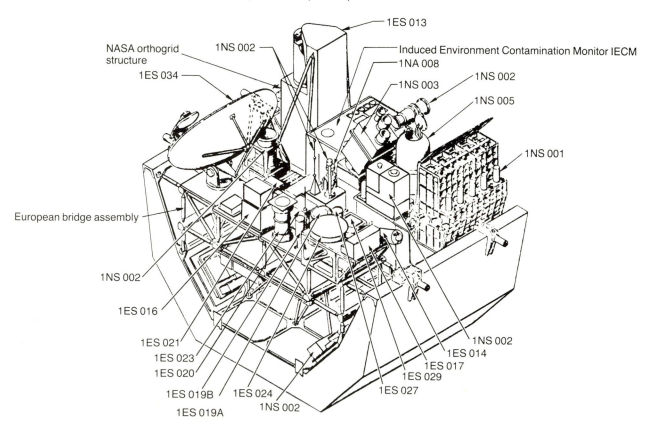

1ES 013

NASA orthogrid structure

1NS 002

Induced Environment Contamination Monitor IECM

1ES 034

1NA 008

1NS 003

1NS 002

1NS 005

1NS 001

European bridge assembly

1NS 002

1ES 016

1ES 021

1ES 023

1ES 020

1ES 019B    1ES 024

1ES 019A    1NS 002

1NS 002

1ES 014

1ES 017

1ES 029

1ES 027

## Spacelab one module experiments — starboard side

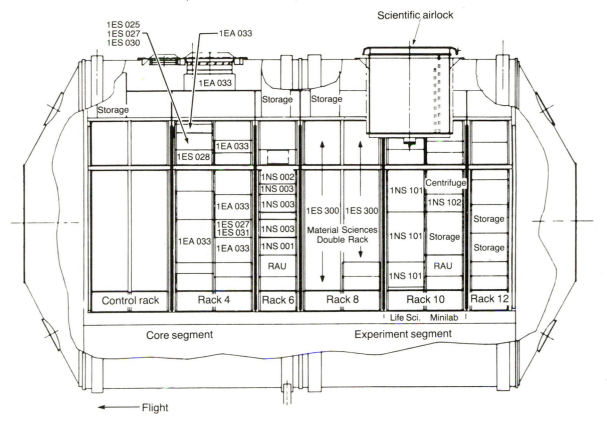

## Spacelab one module experiments – port side

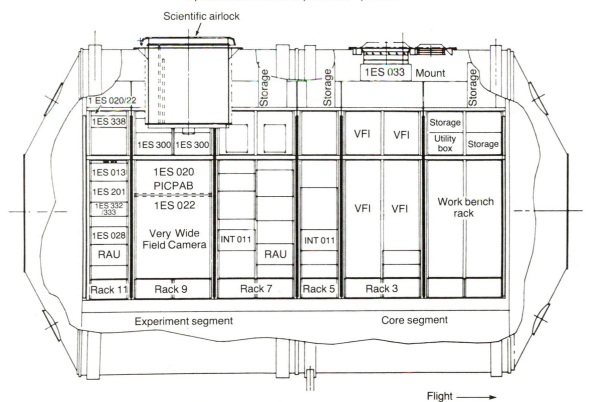

# LIST OF ACRONYMS

| | |
|---|---|
| AC | alternating current |
| AEPI | Atmospheric Emission Photometric Imaging |
| BRS | Body Restraint System |
| CDMS | Command and Data Management Subsystem |
| CHASE | Coronal Helium Abundance Spacelab Experiment |
| CITE | Cargo Integration Test Equipment |
| CNES | Centre National d'Etudes Spatiales |
| CPU | Central Processing Unit |
| DC | direct current |
| DDU | Data Display Unit |
| DFVLR | Deutsche Forschung und Versuchsanstalt für Luft- und Raumfahrt |
| DPA | Data Processing Assembly |
| ECAS | Experiment Computer Application Software |
| ECG | electrocardiograph, electrocardiogram |
| ECOS | Experiment Computer Operating System |
| ECS | Environmental Control Subsystem |
| ECS | European Communications Satellite |
| EEG | electroencephalogram |
| ELDO | European Launcher Development Organisation |
| EM | Spacelab Engineering Model |
| EOG | electro-oculogram |
| EOM-1 | First Earth Observation Mission |
| EPDB | Experiment Power Distribution Box |
| EPDS | Electrical Power Distribution Subsystem |
| EPSP | Experiment Power Switching Panel |
| ESA | European Space Agency |
| ESRO | European Space Research Organisation |
| ESTEC | ESA's technology centre |
| EURECA | European Retrievable Carrier |
| EVA | Extra-Vehicular Activity |
| FAUST | Far Ultra-violet Space Telescope |
| FSLP | First Spacelab Payload |
| FU | Spacelab Flight Unit |
| GAS | Get Away Specials |
| GIRL | German Infra-Red Laboratory |
| GMT | Greenwich Mean Time |
| GN&C | Guidance, Navigation and Control computer |
| GSE | Ground Support Equipment |
| HDRR | High Data Rate Recorder |
| HOSC | Huntsville Operations Support Center |
| HRDA | High Rate Data Assembly |
| IECM | Induced Environment Contamination Monitor |
| IML | International Microgravity Laboratory |
| IPS | Instrument Pointing System |
| IRAS | Infra-Red Astronomical Satellite |
| ISO | Imaging Spectrometric Observatory |
| IUS | Inertial Upper Stage |
| IWG | Investigators' Working Group |
| JSC | Johnson Space Center |
| JSLWG | Joint Spacelab Working Group |
| KSC | Kennedy Space Center |
| LFC | Large Format Camera |
| MAU | million accounting units |
| MEA | Material Experiment Assembly |
| MEDEA | Materials Laboratory |
| MET | mission elapsed time |
| MIT | Massachusetts Institute of Technology |
| MMU | Manned Maneuvering Unit |
| MMU | Mass Memory Unit |
| MOCR | Mission Operations Control Room |
| MOMS | Modular Opto-electronic Multi-spectral Scanner |
| MOU | Memorandum of Understanding |
| MRSE | Microwave Remote Sensing Experiment |
| MSDR | Material Sciences Double Rack |
| MSFC | Marshall Space Flight Center |
| MSL | Material Sciences Laboratory |
| NASA | National Aeronautics and Space Administration |
| NAVEX | navigation experiments |
| NOAA | National Oceanic and Atmospheric Administration |
| O&C | Operations & Check-out Building |
| OCS | Orbiter Control System |
| OMS | Orbital Maneuvering System |
| OPF | Orbiter Processing Facility |
| OSS | NASA Office of Space Sciences |
| OSTA | NASA Office of Space and Terrestrial Applications |
| OTV | Orbital Transfer Vehicle |
| PAM | Payload Assist Module |
| PDP | Plasma Diagnostics Package |
| PICPAB | Phenomena Induced by Charged Particle Beams |
| POCC | Payload Operations Control Center |
| RAU | Remote Acquisition Unit |
| RCS | Reaction Control System |
| RMS | Remote Manipulator System |
| SAR | Synthetic Aperture Radar |
| SBS | Satellite Business System |
| SEPAC | Space Experiments with Particle Accelerators |
| SHEAL | Shuttle High Energy Astrophysics Laboratory |
| SIR | Shuttle Imaging Radar |
| SIRTF | Shuttle Infra-Red Telescope Facility |
| SL | Spacelab |
| SOT | Solar Optical Telescope |
| SPAS | Shuttle Pallet Satellite |
| SPICE | Spacelab Payload Integration and Coordination in Europe |
| SRB | Solid-fuel Rocket Booster |
| SRL | Shuttle Radar Laboratory |
| STDN | Space Tracking and Data Network |
| STS | Space Transportation System |
| TDRS | Tracking and Data Relay Satellite |
| TDRSS | Tracking and Data Relay Satellite System |
| TMS | Teleoperator Maneuvering System |
| UHV | Ultra High Vacuum |
| VAB | Vehicle Assembly Building |
| VFI | Verification Flight Instrumentation |
| VFT | Verification Flight Test |

# INDEX

Page numbers in italic type indicate figures.